Ionization Potentials

Some Variations,
Implications and Applications

Related Pergamon Titles of Interest

Books

AHRENS
Origin and Distribution of the Elements — Proceedings of the 2nd Symposium

HENDERSON
Inorganic Geochemistry

HÖGFELDT
Stability Constants of Metal–Ion Complexes, Part A: Inorganic Ligands

MEITES
An Introduction to Chemical Equilibrium and Kinetics

PERRIN
Stability Constants of Metal–Ion Complexes, Part B: Organic Ligands
Ionisation Constants of Inorganic Acids and Bases in Aqueous Solution

WHITTAKER
Crystallography: An Introduction for Earth Science (and Other Solid State) Students

Journals

Chemistry International
Geochimica et Cosmochimica Acta
Physics and Chemistry of the Earth
Polyhedron
Progress in Solid State Chemistry

Full details of all Pergamon publications/free specimen copy of any Pergamon journal available on request from your nearest Pergamon office.

Ionization Potentials

Some Variations, Implications and Applications

BY

L. H. AHRENS

Emeritus Professor and Senior Research Fellow, Department of Geochemistry, University of Cape Town

PERGAMON PRESS

OXFORD · NEW YORK · TORONTO · SYDNEY · PARIS · FRANKFURT

U.K.	Pergamon Press Ltd., Headington Hill Hall, Oxford OX3 0BW, England
U.S.A.	Pergamon Press Inc., Maxwell House, Fairview Park, Elmsford, New York 10523, U.S.A.
CANADA	Pergamon Press Canada Ltd., Suite 104, 150 Consumers Road, Willowdale, Ontario M2J 1P9, Canada
AUSTRALIA	Pergamon Press (Aust.) Pty. Ltd., P.O. Box 544, Potts Point, N.S.W. 2011, Australia
FRANCE	Pergamon Press SARL, 24 rue des Ecoles, 75240 Paris, Cedex 05, France
FEDERAL REPUBLIC OF GERMANY	Pergamon Press GmbH, Hammerweg 6, D-6242 Kronberg-Taunus, Federal Republic of Germany

First edition 1983

Library of Congress Cataloging in Publication Data

Ahrens, L. H. (Louis Herman), 1918-
Ionization potentials.
Includes indexes.
1. Ionization. I. Title.
QD561.A33 1983 541.3'722 82-23970

British Library Cataloguing in Publication Data

Ahrens, L.H.
Ionization potentials.
1. Ionization—Tables
I. Title
547.1'3722 QC702

ISBN 0-08-025274-5

In order to make this volume available as economically and as rapidly as possible the author's typescript has been reproduced in its original form. This method unfortunately has its typographical limitations but it is hoped that they in no way distract the reader.

Printed in Great Britain by A. Wheaton & Co. Ltd

This book is dedicated to my wife, with much appreciation, and my children Yolande, Wendy and Ian.

Acknowledgements

My acknowledgements go to many, in particular Associate Professor M.J. Orren for his continued encouragement and assistance and to Professor A.J. Erlank for encouragement and interest and D.H. Thornton for his interest and assistance.
Mrs. G. Verblun typed the M.S. and the diagrams were prepared by Mrs. Suzan Abraham. Dr. A.B. Simpson assisted in some of the calculations. The above mentioned are all of the University of Cape Town.
Finally I wish to acknowledge the ready co-operation of staff members of Pergamon Press.

Contents

Introduction

"Our business is with the causes of sensible effects." Newton.

Several parameters of the atom determine the physical and chemical properties of the elements. The ionization potential is one of these and may be defined as the energy required to remove an electron from an atom or ion. The term, ionization energy, is sometimes used in place of ionization potential.

The ionization potential is a uniquely important fundamental property and its applications are wide and varied. In the field of physics, ionization potential data clearly support the existence of shells, half-shells and sub-shells of electrons and distinctions may be made between *s*,*p* and *d* electrons; screening (shielding) may be understood and defined, and, arising from this and observations in isoelectronic sequences, the effective nuclear charge may also be defined with reasonable confidence. A consequence is our understanding of the Periodic Table. As far back as 1950, Finkelnburg states in his book, *Atomic Physics*: "According to Weizel the electron-shell structure, and with it the periodic table, can be more clearly and quantitatively understood by making use of the empirical ionization energies. This also leads to a deeper understanding of the behaviour of the atomic electrons."

Other statements relating to the importance of the ionization potential, particularly in the field of physics, appear in several papers published by Lakatos and his co-workers. One paper, Lakatos, Bohus and Medgyesi (1959), gives several references and contains the following comments: "It is known from our investigations that the *main factors* determining the physical and chemical properties of the elements are: (a) the *effective principal quantum number* (and, respectively the size (*r*) proportional to it) of the valence electron shell of atomic or ionic constituents, (b) the *effective nuclear charge number* of the constituents, i.e. the value obtained by subtracting the *screening* and, respectively, the *shielding number* of electrons from the positive charge number of the nucleus."

Much of what is stated above relates to the use of the ionization potential.

It is in chemistry, more so than any other discipline, that the ionization potential appears to have its widest application. Experience in the teaching of chemistry has indicated to me that the property is particularly useful when attempting to understand and gain an insight into chemical behaviour, notably that of the metals. Examples include: oxidation (valence) states of the metals; chemical inertness; nature of the chemical bond; formation and stabilities of metal-organic complexes; heats of hydration; and, structures of some inorganic compounds. In *geo*chemistry too, application is wide, particularly when we attempt to understand the general problem of distribution of the elements in the minerals, rocks, sediments and waters of our planet. The ionization potential appears to be useful also in *bio*chemistry, or perhaps more correctly, biological systems; for example, when considering the importance of metals with regard to their binding to amino acids, proteins and other organic molecules of biological significance.

While on the discipline of chemistry and associated fields, a few comments will be made on the basic attitude and approach which has been adopted in writing this book. Considerable emphasis will be placed on forces and, as far as possible, physical reality. The writer was once asked to give an inaugural lecture which he entitled, "Physical reality or mathematical quackery?" The title was purposely chosen to be provocative and stimulate sceptical enquiry and was not intended to reduce the importance of mathematics - not at all - but rather to emphasize the need to continue to strive to understand, even if only in small part, the physical reality of whatever the problem happens to be. Consider, for example, the role that resonance and hybridization has played in the development of chemistry. Common usage of this concept over a long period has lulled many a scientist into a state of mind that the implied phenomena are physically real because of the elegance of the mathematical treatment.

Some mathematicians are partly aware of the dangers of the mathematical theory of nature and Poincaré stated as early as 1929 (Poincaré, 1929): "The mathematical method, by its apparent rigour and inflexible course, often inspires in us a confidence nothing warrants and prevents our looking about us."

It is perhaps appropriate to give a little thought on the use of energy with regard to the purpose of this book and I wish to quote from Brown (1956): "Energy is not a thing, or an interaction between things, but what it is defined to be, a metrical feature of a physical process or state. Energy, therefore, can never be a *causal* agent in science; it can never take the place of *force*, and its increasing use as a causal agent only goes to justify Spengler's description of it as the great myth of Western science."

Although the focus in this book is on ionization potentials, other properties and concepts will in addition have to be taken into account when attempting to understand more fully several of the aspects and phenomena which are to be considered. It is realized that although the ionization potential of a molecule is an important parameter, only atomic and ionic species will be considered.

The ionization potential is a highly significant parameter for controlling the properties of electric discharges, in particular those which are thermal and arc-like, where temperature is a vital criterion for determining properties of the discharge. Important as they are, such applications of the ionization potential are considered to be outside the scope of this book. One example of its application is that described by the writer (Ahrens, 1950; Ahrens and Taylor, 1961) for the chemical (qualitative and quantitative) analysis of minerals, rocks, meteorites and lunar materials. In his book dealing with excitation sources, Boumans (1966) makes considerable use of the ionization potential.

THE DATA

The literature of ionization potentials is vast and extends back to before 1920. Most estimates are obtained spectroscopically by determining the convergence limits in the appropriate spectral series and many of these are known with a high degree of accuracy. If it is not possible to determine convergence limits accurately, other means, including interpolation and extrapolation, have been used. Some such estimates are reasonably accurate but most are not satisfactory for the purpose of this book.

Several ionization potential compilations have been made extending back to the 1930s. A publication by Kizer (1961) provides an indication of the amount of work that has gone into obtaining ionization potentials and more than 500 references are listed. For some elements, numerous measurements have been made in order to refine earlier ones and in his exhaustive study, Kizer gives twenty-seven values for the first ionization potential of He, twenty-five values for the first ionization potential of Ne and twenty-six for the first ionization potential of Ar. For each element the earliest measurements were made in 1917.

The compilation that will be used here is that of Moore (1971) except for the Rare Earths for which the data of Martin *et al.* (1974) will be used.

The data of Moore are listed in the Appendix and those of Martin *et al.* in Chapter 5.

The two most commonly used units for expressing ionization potentials are the electron volt (eV) and the wave number (cm^{-1}). The electron volt will be used throughout this book (1 eV = 8066 cm^{-1}, or 1 cm^{-1} = 0.000 012 398 eV).

The contents of Chapters 13, 14 and 15 are with minor additions and modifications reproductions of original papers. Rather than tamper with the contents of these papers it was decided to reproduce them in their entirety. Some repetition will become apparent and it should be noted that for these three chapters the bibliography is included within the chapter and not at the end of the book. The author understands that by adopting the above procedure he is open to criticism. Nevertheless, he feels that a rewrite might lower the possible significance of these papers and that they should therefore be reproduced in full.

LITERATURE COVERAGE

The literature on ionization potentials is vast indeed and it would be a near impossible task to provide adequate coverage, not only because of the sheer volume of the literature but also because of the wide spread of topics that are involved. As a consequence, literature coverage has been selective, but hopefully only a relatively few publications of significance have been omitted.

CHAPTER 1

First Ionization Potentials, Some Major Trends and Features

The first ionization potentials of the elements are related to atomic number in Fig. 1.1. The major shells, K, L, M, N, O, P and Q are indicated as well as the principal quantum number, n. Several trends, discontinuities and other features are apparent and some of these will be considered in turn.

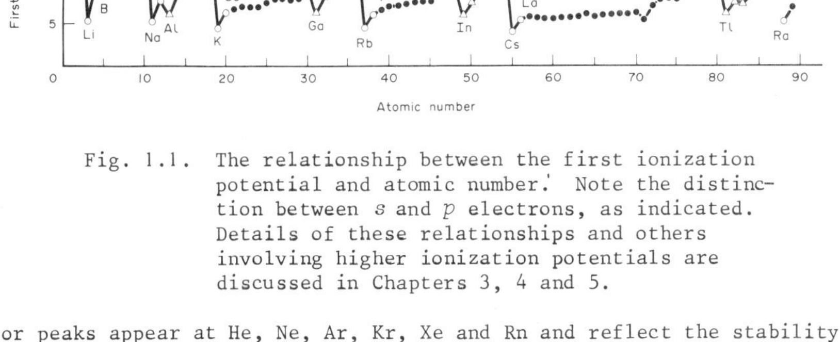

Fig. 1.1. The relationship between the first ionization potential and atomic number. Note the distinction between s and p electrons, as indicated. Details of these relationships and others involving higher ionization potentials are discussed in Chapters 3, 4 and 5.

Major peaks appear at He, Ne, Ar, Kr, Xe and Rn and reflect the stability of the s^2 pair and the s^2p^6 octet, as much energy is required to remove the outermost s electron from He and the outermost p electrons from Ne, Ar, Kr, Xe and Rn. Minima are distinct at Li, Na, K, Rb and Cs as the single s electrons from outside the s^2 pair and the s^2p^6 octets are easily removable.

A few first ionization potentials are given in Table 1.1. These refer to the alkali metals and the noble metals. The contrast in the magnitudes of the ionization potentials is marked and provide a quantitative comparison of the features noted above.

TABLE 1.1

The first *IP*s of the alkali metals and the noble gases

Li	5.39 eV	He	24.59 eV
Na	5.14 "	Ne	21.56 "
K	4.34 "	Ar	15.76 "
Rb	4.18 "	Kr	14.00 "
Cs	3.89 "	Xe	12.13 "

Upward trends beginning with an alkali metal are clearly evident for the short periods Li to Ne and Na to Ar. Similar upward trends involve the elements Cu to Kr, Ag to Xe and Au to Rn. Within each such upward trend discontinuities may be recognized, some distinct and others less so. Such discontinuities are due to sub-shell and half-shell effects.

The sub-shell effect is at a maximum in the first short period and decreases progressively with atomic number: this feature is clearly evident in Fig. 1.1.

The half-shell effects evident in Fig. 1.1 involve *p* electrons only. Such effects decrease in magnitude with atomic number and are quantitatively examined in Chapter 3 which is devoted to *p* electrons only and where both first and higher ionization potentials are considered.

The effects of half-shell closure of the *d* and *f* transition metals are barely in evidence, if at all, in Fig. 1.1. Such effects do exist provided higher ionization potentials as well as the first are considered. The *d* transition elements are discussed in Chapter 4 and the Rare Earths in Chapter 5.

A small but distinct elevation at Hg is apparent in Fig. 1.1. This high *IP* may have a significant bearing on the chemical behaviour of Hg (Ahrens, in prep.).

The maxima and minima in Fig. 1.1 become less evident as atomic number increases and such decreases are clear in Table 1.1. The decrease of ionization potential within each group is regular and clear. Such trends are typical within each group in the Periodic Table and are observable in Fig. 1.1. Quantitative comparisons can be made by referring to the Table of Ionization Potentials given in the Appendix.

Although the above trends are typical, they may be disturbed, but in varying degrees, by the entry of the 4*f* electrons, a feature which is often striking and can have significant consequences on the properties of the elements within one group. The Cu, Ag, Au relationship is one such example (p.43) and others are referred to on pp.47,50. Ahrens (in prep.) has discussed in detail the effects which closure of the 4*f* shell has had on a variety of properties of the elements.

Much of the purpose of this chapter has been to provide an overall view of the relationship between the first *IP* and atomic number. In Chapters 2, 3, 4 and 5 each electron type is discussed quantitatively in some detail and both first and higher ionization potentials, where available, will be considered.

CHAPTER 2

The *s* Electrons

THE RELATIONSHIP BETWEEN SUCCESSIVE *IP*s

Pyper and Grant (1978) have discussed successive *IP*s mainly of the *s* electrons. They provide a theoretical basis for their conclusions on such relationships and their paper should be consulted by those interested in a theoretical discussion.

The data to be considered here are given in Tables 2.1, 2.2, 2.3, 2.4 and 2.5.

Two groups of elements will be considered in the discussion which follows:

(i) The Group 2 elements.

(ii) Several of the *d*-transition elements and all of the *f* transition elements.

(i) The differences between the successive *IP*s of the Group 2 elements are relatively small (5-10 eV) and all the ratios are somewhat less than 2; the average for the Group 2 (a) elements (1.95) is greater than that of the Group 2 (b) elements (1.86).

(ii) With the exception of Sc (1.95) and Ti (2.00) all ratios of the 3*d*-transition elements are greater than 2. Most ratios of the 4*d*-transition elements are also greater than 2. It should be noted that unlike the Group 2 and Rare Earth elements, all of which have two outermost *s* electrons, some of the *d*-transition elements have only one outer electron and one, Pd, has none.

Differences and ratios involving the Rare Earths are remarkably uniform with the exception of Y and Lu. The average ratio of 1.96 is virtually identical to the average for the alkaline earths.

THE RELATIONSHIP BETWEEN THE FIRST *IP*s OF THE GROUP 1 AND GROUP 2 ELEMENTS

The first *IP*s of the Group 1 and Group 2 elements together with the ratios, Group 2/Group 1, are listed in Tables 2.1 to 2.5.

The main features are, firstly, a regular decrease of the alkaline earth/alkali metal ratio with increase in principal quantum number and second(ly) a distinct discontinuity involving the ratio Hg/Au. This discontinuity is due to the conspicuously high *IP* of Hg (p.98).

TABLE 2.1

First and second *IP*s, differences and ratios.
The Group 2 elements

		I_1(eV)	I_2(eV)	Difference	Ratio
s^2	Be	9.32	18.21	8.89	1.95
s^2	Mg	7.65	15.04	7.39	1.97
s^2	Ca	6.11	11.87	5.76	1.95
s^2	Sr	5.70	11.03	5.33	1.94
s^2	Ba	5.21	10.00	4.79	1.92
				Ave.	1.95
s^2	Zn	9.39	17.96	8.57	1.91
s^2	Cd	9.00	16.90	7.90	1.88
s^2	Hg	10.44	18.75	8.31	1.80
				Ave.	1.86

TABLE 2.2

First and second *IP*s, differences and ratios.
The 3*d* transition elements

		I_1(eV)	I_2(eV)	Difference	Ratio
d^1s^2	Sc	6.54	12.80	6.26	1.95
d^2s^2	Ti	6.82	13.58	6.76	2.00
d^3s^2	V	6.74	14.65	7.91	2.17
d^5s^1	(Cr)	6.77	16.50	9.73	2.44
d^5s^2	Mn	7.44	15.64	8.20	2.10
d^6s^2	Fe	7.87	16.18	8.31	2.06
d^7s^2	Co	7.86	17.06	9.20	2.16
d^8s^2	Ni	7.64	18.17	10.53	2.38
$d^{10}s^1$	(Cu)	7.73	20.29	12.56	2.64

TABLE 2.3

First and second *IP*s, differences and ratios.
The 4*d* transition elements

		I_1(eV)	I_2(eV)	Difference	Ratio
d^2s^2	Zr	6.84	13.13	5.53	1.92
d^4s^1	Nb	8.88	14.32	5.45	2.08
d^5s^1	Mo	7.10	16.15	5.49	2.27
d^5s^2	Tc	7.28	15.26	5.75	2.10
d^7s^1	Ru	7.36	16.76	5.68	2.28
d^8s^1	Rh	7.46	18.10	5.65	2.42
d^{10}	Pd	8.34	19.43	6.40	2.33
$d^{10}s^1$	Ag	7.58	21.49	5.43	2.84

TABLE 2.4

First and second *IP*s, differences and ratios.
The Rare Earth elements

	I_1(eV)	I_2(eV)	Difference	Ratio
Y	6.38	12.24		1.91
La	5.58	11.06	5.48	1.98
Ce	5.47	10.85	5.38	1.98
Pr	5.42	10.55	5.13	1.96
Nd	5.49	10.72	5.23	1.95
Pm	5.55	10.90	5.35	1.95
Sm	5.63	11.07	5.44	1.98
Eu	5.67	11.25	5.58	1.98
Gd	6.14	12.10	5.96	1.98
Tb	5.85	11.52	5.67	1.95
Dy	5.93	11.67	5.74	1.98
Ho	6.02	11.80	5.78	1.97
Er	6.10	11.93	5.83	1.95
Tm	6.18	12.05	5.87	1.95
Yb	6.25	12.17	5.92	1.94
Lu	5.43	13.90	8.47	2.57

TABLE 2.5

Principal quantum number	Element	*IP*(eV)	Ratio
1	H	13.59	
	He	24.59	1.81
2	Li	5.39	
	Be	9.32	1.73
3	Na	5.14	
	Mg	7.65	1.49
4	K	4.34	
	Ca	6.11	1.41
5	Rb	4.18	
	Sr	5.70	1.36
6	Cs	3.89	
	Ba	5.21	1.34
4	Cu	7.7	
	Zn	9.39	1.22
5	Ag	7.58	
	Cd	8.00	1.06
6	Au	9.23	
	Hg	10.44	1.13

CHAPTER 3

The *p* Electrons

In this chapter on the *P* electrons, relationships involving the first *IP* will be considered first followed by relationships involving higher *IP*s.

THE FIRST IONIZATION POTENTIAL

Elements which contain an outer *p* electron are:

Principal quantum number							
2	B,	C,	N	|	O,	F,	Ne
3	Al,	Si,	P	|	S,	Cl,	Ar
4	Ga,	Ge,	As	|	Se,	Br,	Kr
5	In,	Sn,	Sb	|	Te,	I,	Xe
6	Tl,	Pb,	Bi	|	Po,	At,	Rn

The dashed line distinguishes the half-shells.

Figures 3.1 to 3.5 relate the first *IP* to atomic number for each of the elements listed above and as in other Figures in this chapter *IP* differences are shown.

The first three Figures exhibit several well-developed features. For each half-shell the *Z-IP* relationship is almost linear and the slope for the first half-shell is slightly less than that of the second half. Half-shell effects are clear and there is a progressive decrease in the half-shell decrease as follows: N–O, -0.916 eV, P-S, -0.126 eV and As-Se, -0.058 eV.

For the shell, In to Xe, the features noted above are not well defined. *IP* increases more or less regularly with *Z* and a half-shell effect, Sb-Te, is barely perceptible.

The Z-IP relationship for $n = 6$ differs distinctly from the others, notably because the IP for Bi is actually less and not greater than that of Pb; the IP of Bi may therefore be regarded as an anomalous feature which has been discussed by Ahrens (in prep.).

Quantitative data on IP differences for each shell of p electrons for $n = 2$ to $n = 4$ are given in Table 3.1.

TABLE 3.1

IP differences within each half-shell

$n = 2$	C-B	2.96 eV
	N-C	3.27 "
	F-O	3.80 "
	Ne-F	4.12 "
$n = 3$	Si-Al	2.17 "
	P-Si	2.34 "
	Cl-S	2.61 "
	Ar-Cl	2.79 "
$n = 4$	Ge-Ga	1.90 "
	As-Ge	1.91 "
	Br-Se	2.06 "
	Kr-Br	2.19 "

Progressive changes of differences are clear.

For each quantum group, differences for each succeeding pair increase regularly with Z and there is a progressive change with n, from $n = 2$ to $n = 4$.

Close inspection of the data reveals a curious coincidence which might be significant. The average of the two differences, Si-Al and P-Si is exactly 2.25 eV and the sum 4.5 eV. These values coincide exactly with the matching pair average and quarter-shell differences involving d electrons (Chapter 4).

HIGHER IONIZATION POTENTIALS

For the higher *IP*s, *IP* differences are indicated in the various diagrams as in Figs. 3.1 to 3.5 but the differences will not be tabulated.

Figures 3.6 and 3.7 relate the second *IP* to Z for elements of the various quantum groups. The half-shell effects are always positive, unlike the decreases of the first *IP* for $n = 2$, 3 and 4.

For the groups C to Na and Si to K, the relationships are clearly defined, but less so for the groups Ge to Rb and Sn to Cs.

Relationships involving the third, fourth and fifth *IP*s are shown in Figs. 3.7 to 3.12. The relationships are all well developed and several progressive changes are clearly apparent.

For the relatively light elements accurate *IP* data are available up to the fifteenth state and *Z-IP* relationships such as those above persist to this high state of ionization. These relationships are highly quantitative and will be described elsewhere.

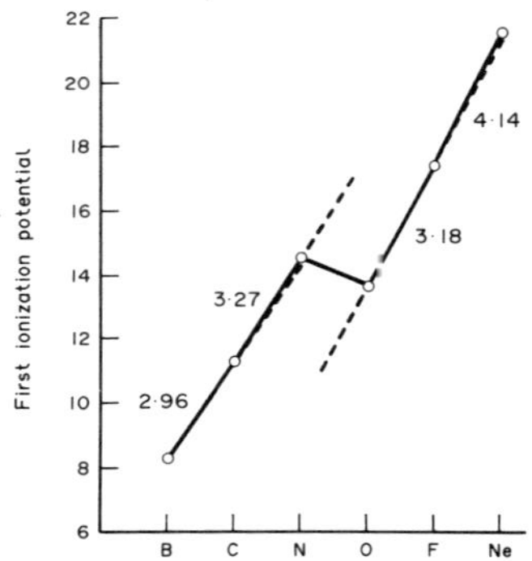

Fig. 3.1. The relationship between the first ionization potential for elements B to Ne. The relationships are almost linear as in many of the subsequent ones. *IP* differences are indicated. The effect of a half-shell is clear and the displacement is similar to those shown in most relationships displayed in Figs. 3.2–3.12.

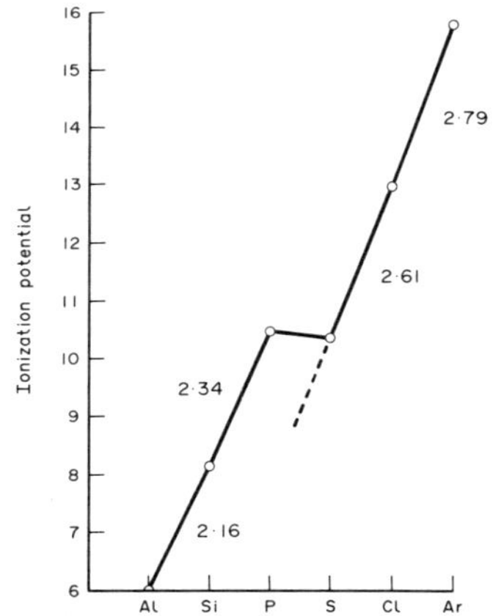

Fig. 3.2. The relationship between ionization potential and the elements Al to Ar.

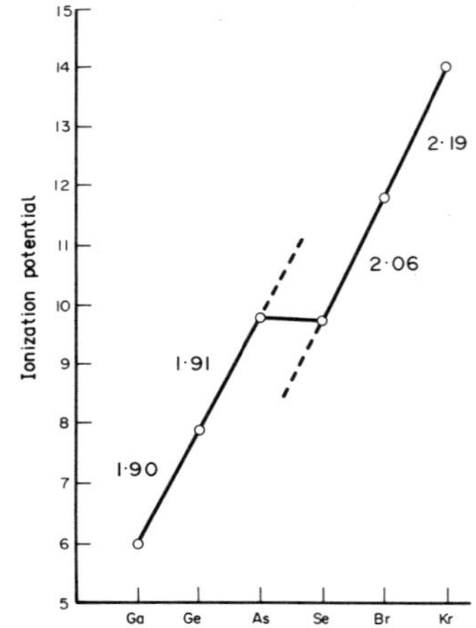

Fig. 3.3. The relationship between ionization potential and the elements Ga to Kr.

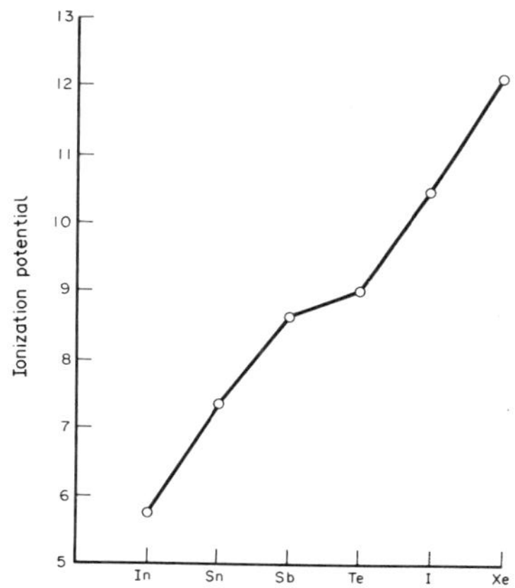

Fig. 3.4. The relationship between ionization potential and the elements In to Xe. Unlike the relationship in Figs. 3.1–3.3 inclusive, this relationship is not well developed. The half-shell displacement is barely evident.

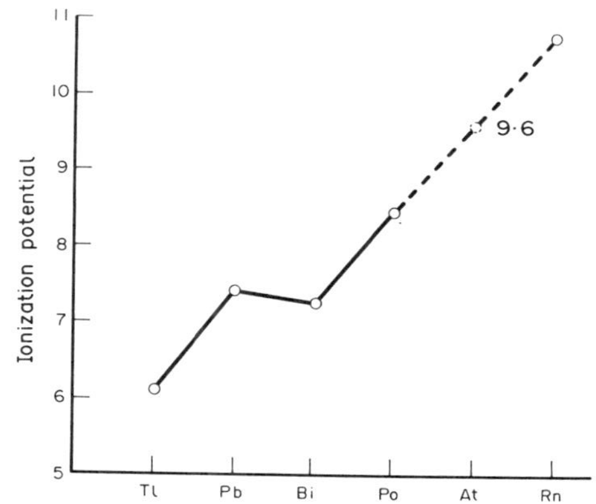

Fig. 3.5. The relationship between ionization potential and the elements Tl to Rn. This relationship is distinctly different from all others involving the first *IP*. No half-shell displacement is evident.

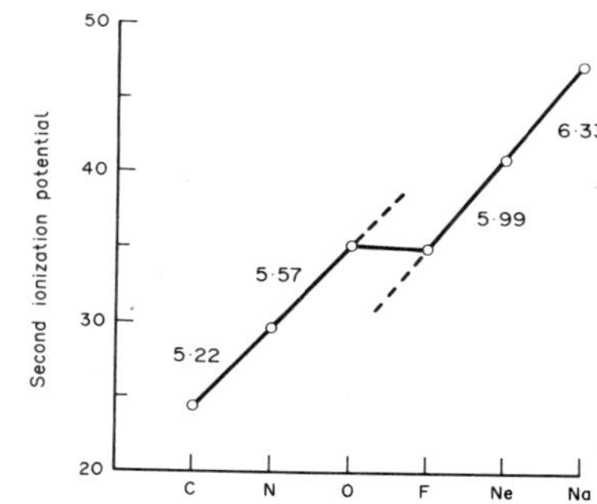

Fig. 3.6. In this diagram involving the second ionization potential from C to Na the features are similar to some of those involving the first *IP*.

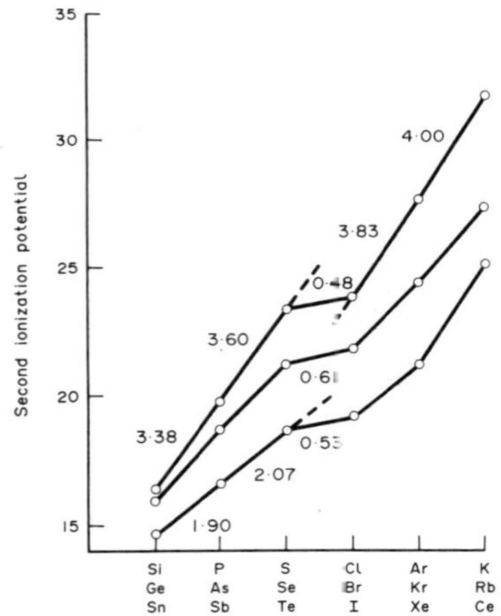

Fig. 3.7. The second ionization potential is related to three sets of elements and it is clear that the Si and K relationship is well developed compared with elements in the other two sequences.

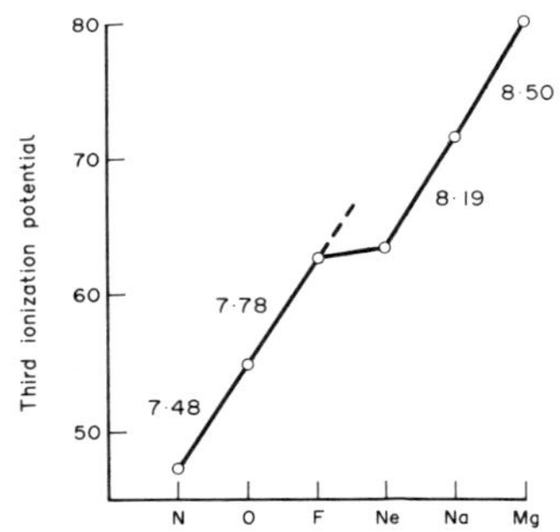

Fig. 3.8. The third ionization potential is related to elements N to Mg and as in all the following diagrams, relationships are well developed.

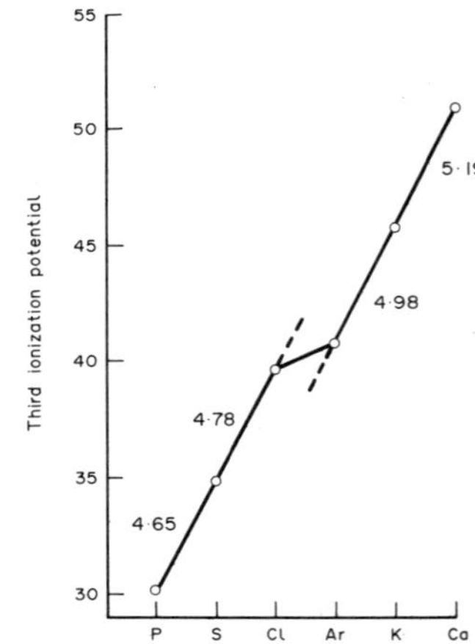

Fig. 3.9. As in Fig. 3.8 but for elements from P to Ca.

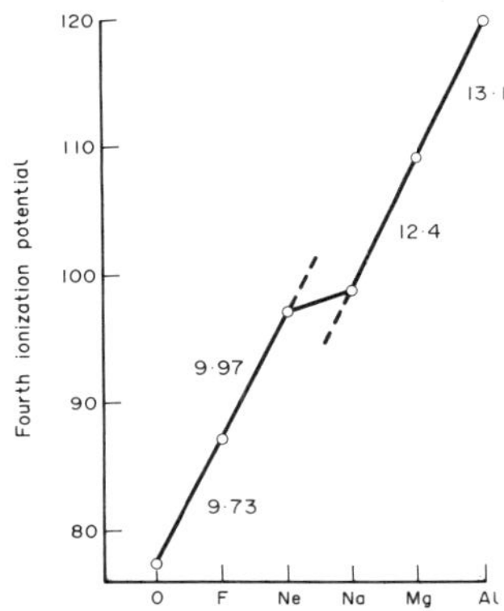

Fig. 3.10. The relationship between the fourth ionization potential and elements O to Al.

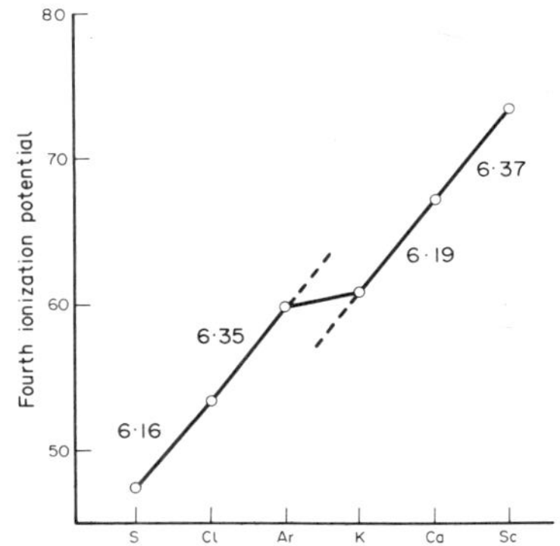

Fig. 3.11. The relationship between the fourth ionization potential and elements S to Sc.

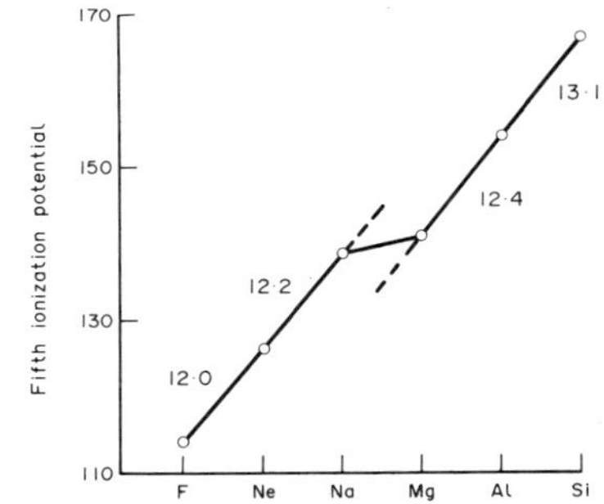

Fig. 3.12. The relationship between the fifth ionization potential and elements F to Si. Relationships involving higher *IP*s have been examined elsewhere and the features in each are well developed and similar to most of those in Figs. 3.1–3.12.

CHAPTER 4

The *d*-Transition Elements

THE FIRST IONIZATION POTENTIAL

Figure 4.1 relates *IP* to *Z* and it is evident that the relationship tends to be irregular, but in a rough and general way *IP* increases with *Z*. For each series, *IP* rises sharply after d^{10} closure (Cu, Ag and Au) to their respective maxima (Zn, Cd and Hg). The 3*d* and 4*d* relationships tend to criss-cross. The reasons are not apparent.

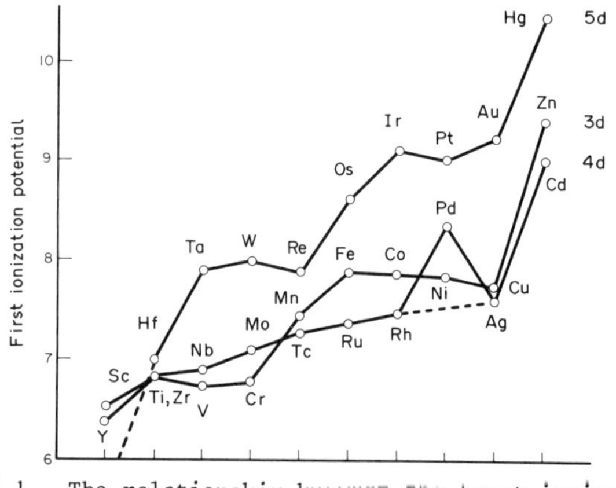

Fig. 4.1 The relationship between the first ionization potential and elements belonging to the 3*d*, 4*d* and 5*d* transition series. It may be noted that the *IP*s for members of the 5*d* transition elements are higher than those of the 3*d* and 4*d* transition elements.

With the exception of Pd, all *d*-transition metals contain either one or two outer *s* electrons, and the elevation at Pd is evidently due to the fact that a *d* electron and not an *s* electron is involved.

The most noteworthy feature involving the first *IP-Z* relationship is the location of the 5*d* series. The *IP*s of each 5*d* element are distinctly greater than the 3*d* and 4*d* counterparts. The minimum difference involves the Group 7 elements, and the maximum, Cu, Ag and Au.

The reason why the *IP*s of the 5*d* elements are relatively high is that these elements follow the Rare Earths and screening by the 4*f* electrons is relatively inefficient (Chapter 8).

Table 4.1 gives half-shell differences for the first and second *IP*s; those of the third *IP*s are tabulated separately.

TABLE 4.1

First and second *IP* half-shell differences

3*d*	Element	*IP*(eV)	Element	*IP*(eV)	Difference
1st *IP*	Sc	6.54	Mn	7.44	0.90
	Fe	7.87	Zn	9.39	1.52
2nd *IP*	Sc	12.80	Mn	15.64	2.84
	Fe	16.18	Zn	17.96	1.78
4*d*					
1st *IP*	Y	6.38	Tc	7.28	0.90
	Ru	7.37	Cd	8.99	1.62
2nd *IP*	Y	12.24	Tc	15.26	3.02
	Ru	16.76	Cd	16.91	0.15

The first half-shell difference (0.90 eV), Sc to Mn, for the 3*d*-transition elements is equal to the first half-shell difference, Y to Tc for the 4*d*-transition elements, and the second half-shell difference (1.52), Fe to Zn, is similar to the second half-shell difference (1.62), Ru to Cd, for the 4*d*-transition elements.

The value of 0.9 eV is exactly one-tenth that of the average half-shell difference of 9.0 eV involving the third *IP* (p. 19).

It is of interest to note that the Sc-Y difference of 0.16 eV is equal to the Mn-Tc difference. The Fe-Ru difference of 0.50 eV is similar to the Zn-Cd difference of 0.40 eV, calculated from Table 4.1.

THE SECOND IONIZATION POTENTIAL

When considering the second *IP-Z* relationship (Fig. 4.2) it should again be borne in mind that some *d*-transition elements contain only one outer *s* electron and for these elements the second *IP* refers to the energy to remove a *d* electron instead of an *s* electron.

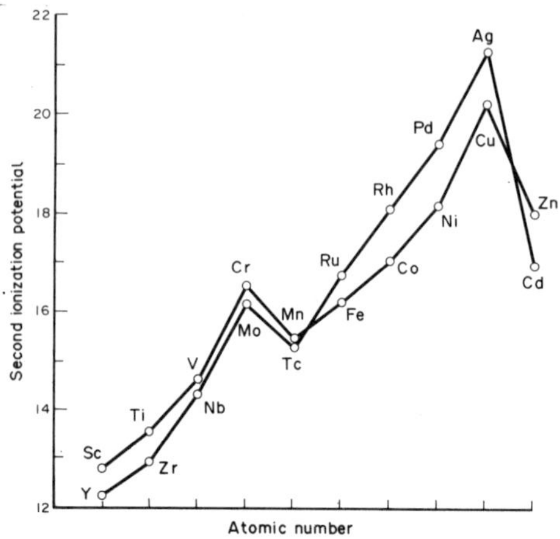

Fig. 4.2. The relationship between the second ionization potential and elements belonging to the $3d$ and $4d$ transition series. Features tend to be more clearly defined than those in Fig. 4.1. A half-shell effect is clear and it is noteworthy that the $3d/4d$ relationship in the first half-shells differs from that in the second half-shell as a result of a crossover at Mn and Tc.

A general upward trend of *IP* with *Z* is clearly apparent in Fig. 4.2 for both $3d$ and $4d$ elements and is punctuated by a half-shell displacement. The cross-over after half-shell closure is of considerable interest as the *IP*s of each $4d$ element is greater than its $3d$ counterpart up to Cu and Ag, after which the $3d$-$4d$ relationship is reversed. The $3d$-$4d$ relationship for the pairs Fe-Ru, Co-Rh and Ni-Pd may be due to the fact each $3d$ element contains two *s* electrons whereas the $4d$ equivalents contain only one. This reasoning does not, however, apply to Cu and Ag as both elements have the same electronic configuration, $d^{10}s^1$.

The first half-shell difference (2.84 eV), Sc to Mn, for the $3d$-transition elements is similar to the first half-shell difference (3.02 eV), Y to Tc, for the $4d$ transition elements. The values for the second half-shells differ completely and this is evidently due to the fact that the electron configuration of Ru is ds^1 and not ds^2 and the second *IP* involves the removal of a *d* electron and not an *s* electron. As a result, the *IP* of Ru is greater than that of Fe and not vice versa, which would have been the case had the configuration of Ru been ds^2.

THE THIRD IONIZATION POTENTIAL

The third ionization potentials of the first and last members of the $3d$, $4d$ and $5d$ shells of the *d*-transition elements are related to atomic number in Fig. 4.3. The slopes of the Sc-Zn and La-Hg joins are parallel and less than that of the Y-Cd join.

Figure 4.4 relates *IP* to *Z* for all 3*d* and 4*d* elements. When preparing this diagram it became apparent that several well-developed relationships could be shown by joining various pairs of elements; these are indicated in Fig. 4.5. For the sake of convenience a letter of the alphabet has been assigned each element.

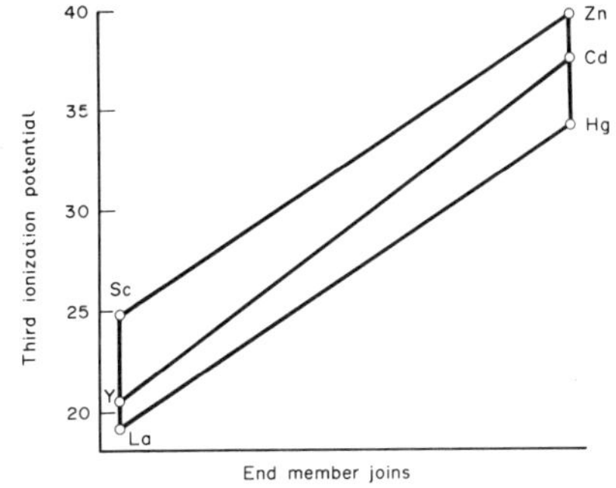

Fig. 4.3. The relationship between the third ionization potential of the end member elements of the 3*d*, 4*d* and 5*d* transition elements.

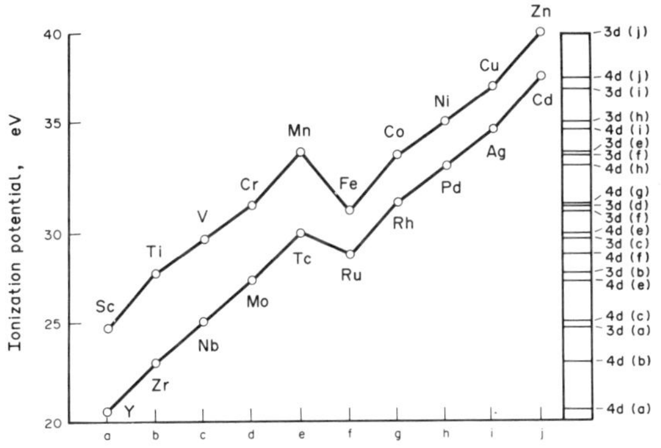

Fig. 4.4. In this diagram involving the third ionization potential, half-shell effects are clear in addition to distinctive trends. The two graphs are similar in form. Letters of the alphabet are indicated in the horizontal scale; see text for details. The vertical column on the right indicates by horizontal lines the respective *IP*s corresponding to each element, as symbolised.

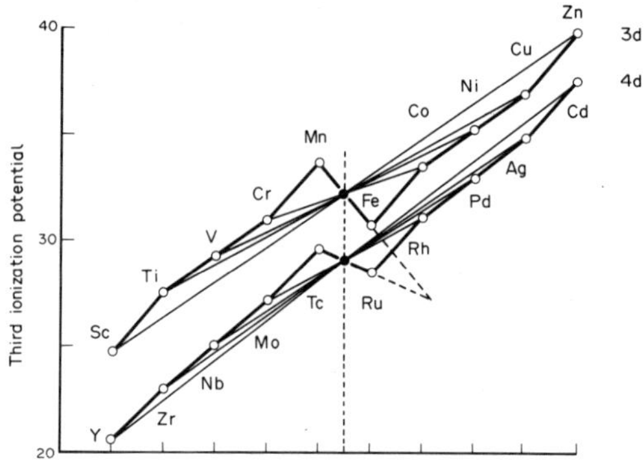

Fig. 4.5. In this diagram various pairs of elements have been joined as indicated. Many striking features, some of them which appear to be of fundamental significance, are detailed in the text.

Several features are clear in Fig. 4.5, and elements may be grouped into half-shells, quarter-shells and matching pairs as shown in Tables 4.2 and 4.3 respectively. The following observations based on Fig. 4.5 and Tables 4.2 and 4.3 may be made:

(i) <u>Half-shells</u>
According to Table 4.2, each half-shell difference is close to 9.0 eV and the average of the four half-shells differences is exactly 9.0 eV.

(ii) <u>Quarter-shells</u>
According to Table 4.2, each quarter-shell difference is close to 4.5 eV and the average of the eight quarter-shell differences is exactly 4.5 eV.

(iii) <u>Matching pairs</u>
Each of the matched pairs in Table 4.3 is either end-member (a, b and d, e and f, g and i, j) or adjacent (b, c and c, d; and g, h and h, i) and each differs from each other.

The average of the sixteen pairs is 2.25 eV. This differs from each of the individual pairs and is half the quarter-shell difference and quarter of the half-shell difference.

(iv) The 3*d*-shell difference (j-a) of 14.96 eV (15.0 eV) is less than the 4*d*-shell difference of 17.0 eV and closely similar to the 5*d*-shell difference of 15.03 eV (14.96 and 15.03 average at 15.0 eV).

Much of what has been written in (i) to (iv) is summarized diagrammatically in Fig. 4.6.

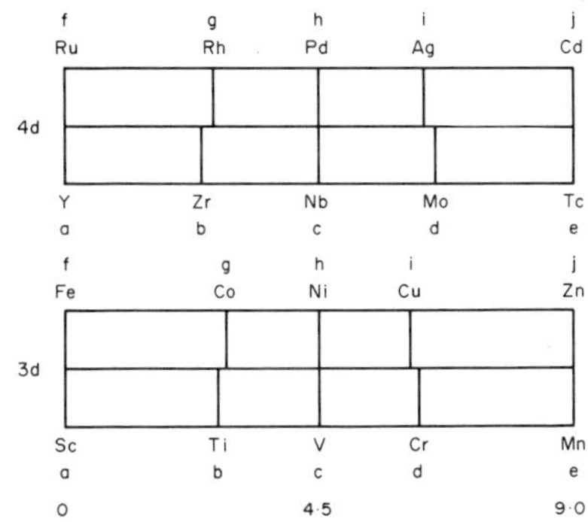

Fig. 4.6. This diagram is intended to summarize the data in Table 4.2. The average half-shell (9.0 eV) and quarter-shell (4.5 eV) differences are indicated as well as the individual matching pair differences.

TABLE 4.2

IP difference *s* involving half-shells and quarter-shells

3*d*	e-a	8.91	4*d*	e-a	9.0
	c-a	4.55		c-a	4.5
	e-c	4.36		e-c	4.5
	j-f	9.07		j-f	9.01
	h-f	4.52		h-f	4.46
	j-h	4.55		j-h	4.55

TABLE 4.3

Matching pairs

3*d*	a, b	b-a = 2.73		
	d, e	e-d = 2.71	Average	2.72
	b, c	c-b = 1.82		
	c, d	d-c = 1.65	Average	1.74
	f, g	f-g = 2.85		
	i, j	j-i = 2.89	Average	2.87
	g, h	h-g = 1.67		
	h, i	i-h = 1.66	Average	1.67
4*d*	a, b	b-a = 2.47		
	d, e	e-d = 2.38	Average	2.43
	b, c	c-d = 2.05		
	c, d	d-c = 2.12	Average	2.09
	f, g	g-f = 2.59		
	i, j	j-i = 2.65	Average	2.62
	g, h	h-g = 1.87		
	h, i	i-h = 1.90	Average	1.89

Overall average 2.25

Figure 4.7 has been obtained from Fig. 4.6 and the data of Tables 4.2 and 4.3. The averaged half-shell and quarter-shell differences are shown as well as the matching pair differences. The averages of each difference (half-shell, quarter-shell and matching pair) are shown in Fig. 4.8.

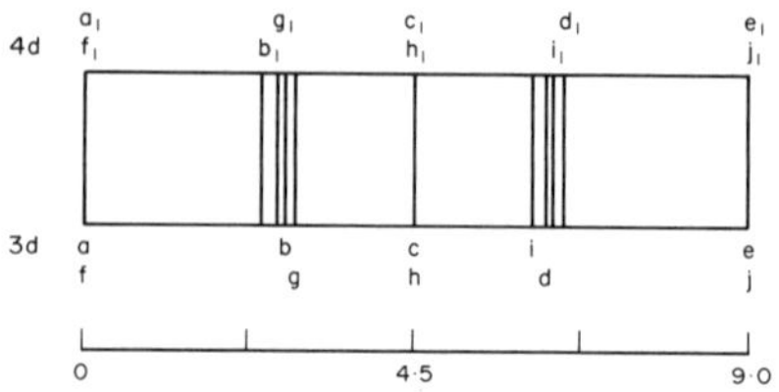

Fig. 4.7. As in Fig. 4.6, but the half-shell and quarter-shell averaged differences have been reduced into one strip, together with the individual matching pair differences. The alphabetical symbols for the 3*d* transition elements have been distinguished from the corresponding 4*d* transition elements.

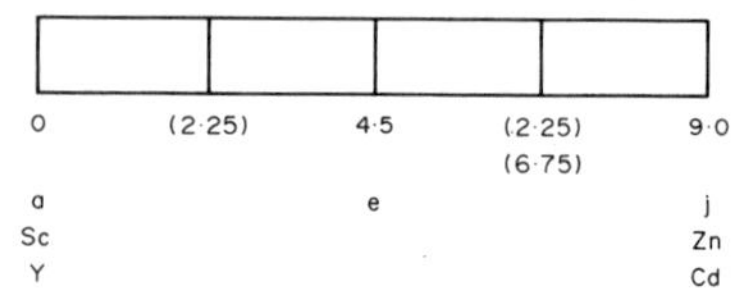

Fig. 4.8. In this simplified diagram averaged values only are indicated. The four-fold division involving a factor of 2.25 eV is clear.

(v) Intersecting joins

For both the 3*d* and 4*d* elements, joins of a,j; b,i; c,h; d,g and e,f (the half-shell join) define a common point. Both these points lie midway between adjacent atomic numbers involved in half-shell closure. It has been proposed (Ahrens, in prep.) to label them *DM* after Dimitri Mendeleev. The respective *Z* and *IP* values for DM_1 (3*d* elements) and DM_2 (4*d* elements) are, 25.5 and 32.2 eV, and 43.5 and 29.0 eV.

As both DM_1 and DM_2 appear to correspond to one-half *Z*, it is possible that that half an e.s.u. may be important in element build-up, at least for the d-transition elements (Ahrens, in prep.).

(vi) The third ionization potentials of all 3*d* transition elements are greater than their 4*d* counterparts, unlike the first and second *IP*s.

(vii) Differences between 3*d* elements and their 4*d* counterparts

The differences between each 3*d* element and its 4*d* counterpart are shown in Table 4.4.

It is clear that for both the 3*d* and 4*d* elements, *IP* varies regularly with *Z* and such a relationship appears to be sinusoidal (Fig. 4.9). The respective curves are similar in form.

TABLE 4.4

IP differences between pairs of elements of the same Periodic Group

Sc, Y	4.24 eV	
Ti, Zr	4.50 "	
V, Nb	4.27 "	
Cr, Mo	3.80 "	
Mn, Tc	4.13 "	Ave. 4.19
Fe, Ru	2.18 "	
Co, Rh	2.44 "	
Ni, Pd	2.24 "	
Cu, Ag	2.00 "	
Zn, Cd	2.24 "	Ave. 2.22

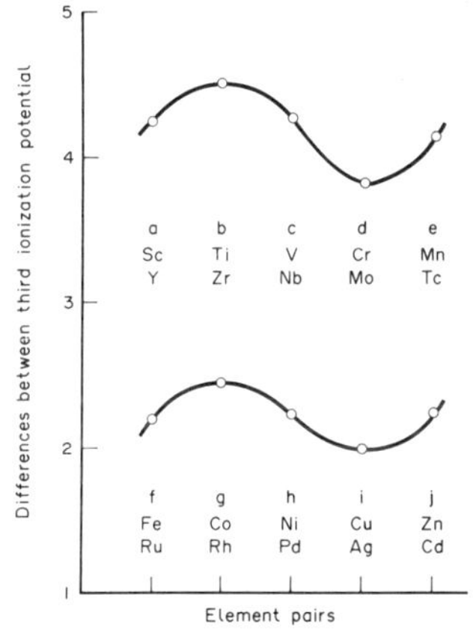

Fig. 4.9. Differences between the third ionization potentials for each corresponding element pair are compared. The patterns are similar but differ quantitatively.

SOME EARLY OBSERVATIONS USING APPROXIMATE DATA

Using relatively approximate data as listed in the Tabellen of Landolf-Börnstein, Ahrens (1957) drew attention to the fact that the difference between the third *IP*s of Zn and Ga was approximately 9 eV. The values of Moore are 39.72 eV for Ga and 30.71 eV for Zn and the difference is therefore 9.01 eV. Ahrens observed further that for successive higher *IP*s for the pairs Ga-Ge (4th *IP*), Ge-As (5th *IP*) up to Rb-Sr (10th *IP*), several of the differences involved multiples of nine. Few of the available high *IP*s are accurate, but it may be noted that for the pair Ga-Ge the respective fourth *IP*s are 64.0 eV and 45.7 *IP* and that the difference of 18.3 eV is close to twice nine. To what extent these correlations may be significant is difficult to ascertain.

THE FOURTH IONIZATION POTENTIAL

The fourth *IP* data are relatively inaccurate. Nevertheless, some trends and features have been observed (Ahrens, in prep.).

CHAPTER 5

The Rare Earth Elements

THE DATA

The data to be used are those of Martin *et al.* (1974). These authors compiled values for the first four *IP*s, each of which was determined spectroscopically and it is clear from their publication that the data they list are of high quality, which is necessary for the purpose at hand. The data are given in Table 5.1.

TABLE 5.1

The first four *IP*s of the Rare Earth elements

Z	Element	I_1(eV)	I_2(eV)	I_3(eV)	I_4(eV)
57	La	5.58	11.06	19.18	49.95
58	Ce	5.47	10.85	20.00	36.76
59	Pr	5.42	10.55	21.62	38.98
60	Nd	5.49	10.72	22.10	40.41
61	Pm	5.55	10.90	22.30	41.10
62	Sm	5.63	11.07	23.40	41.40
63	Eu	5.67	11.25	24.92	42.60
64	Gd	6.14	12.09	20.63	44.00
65	Tb	5.85	11.52	21.91	39.79
66	Dy	5.93	11.67	22.80	41.47
67	Ho	6.02	11.80	22.84	42.50
68	Er	6.10	11.93	22.74	42.65
69	Tm	6.18	12.05	23.68	42.69
70	Yb	6.25	12.18	25.03	43.74
71	Lu	5.43	13.90	20.96	45.19

DISCUSSION

A preliminary examination of the relationship between the first four *IP*s and Z revealed several distinct features. It was observed that those involving the first and second *IP*s were similar but differed from those which involved the third and fourth *IP*s. As a result, the *IP*-*Z* relationship will be considered in two parts; the first deals with the first and second *IP*s and the second with the third and fourth *IP*s.

THE FIRST AND SECOND *IP*s

A scan of the first and second *IP*s in Table 5.1 shows several features, including trends and discontinuities. These are clearly revealed when *IP* is related to *Z* (Figs. 5.1 and 5.2).

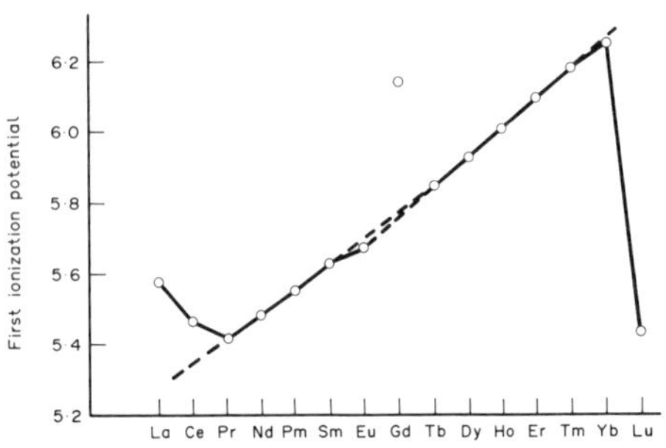

Fig. 5.1. The first ionization potential is related to each of the Rare Earth elements. Several features are clear including a half-shell discontinuity and close correlations for some groups of elements.

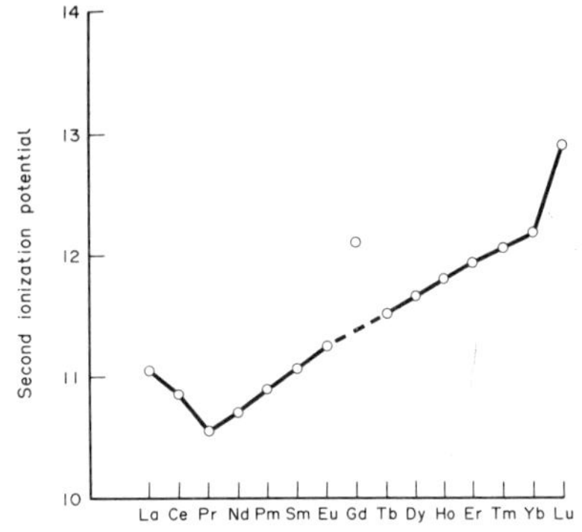

Fig. 5.2. For the second ionization potential a half-shell effect is clear as well as the presence of close correlations.

Three distinct features are evident:

(i) *IP* decreases in the sequence La, Ce and Pr;

(ii) these decreases are followed by regular small increases up to *Z* = 63 (Eu) and then a sharp increase at *Z* = 64 (Gd) which is attributable to half-shell closure of seven 4*f* electrons;

(iii) regular small increases begin again at *Z* = 65 (Tb) and terminate at *Z* = 70 (Yb). *IP* drops sharply at Lu (*Z* = 71) which signals closure of the full shell of fourteen 4*f* electrons.

It should be noted that whereas the half-shell discontinuity involves a sharp *IP* increase, half-shell discontinuities for *p* and *d* electrons usually involve a decrease in *IP* (Chapters 3 and 4).

Several of the regular *IP-Z* increases in Figs. 5.1 and 5.2 appear to be linear and in order to examine this possibility more sensitively, parts of the relationships shown in Figs. 5.2 and 5.3 have been redrawn using enlarged scales (Figs. 5.3, 5.4, 5.5 and 5.6).

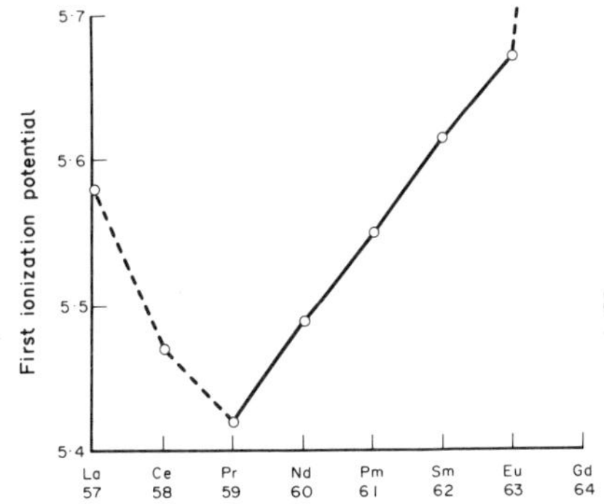

Fig. 5.3. As in the first half of Fig. 5.2 but using expanded scales.

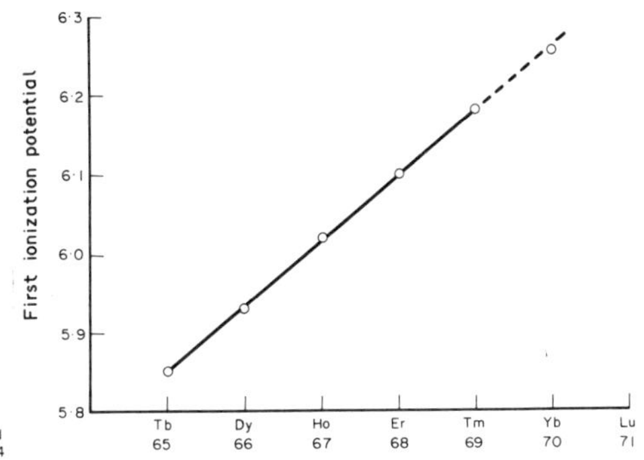

Fig. 5.4. A close correlation involving five elements is clearly evident.

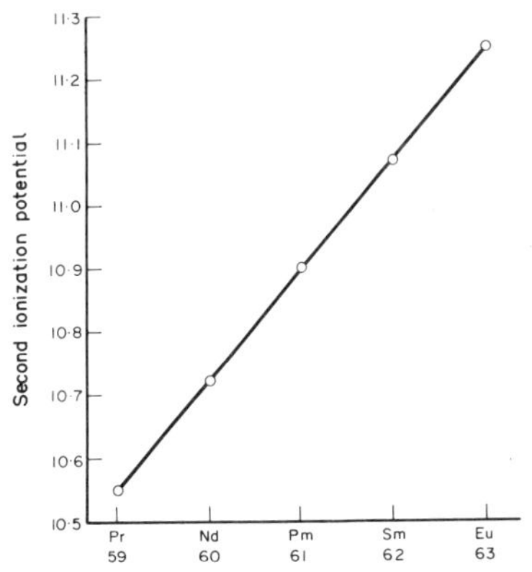

Fig. 5.5. As in Fig. 5.4 correlation is very close indeed and appears to be linear.

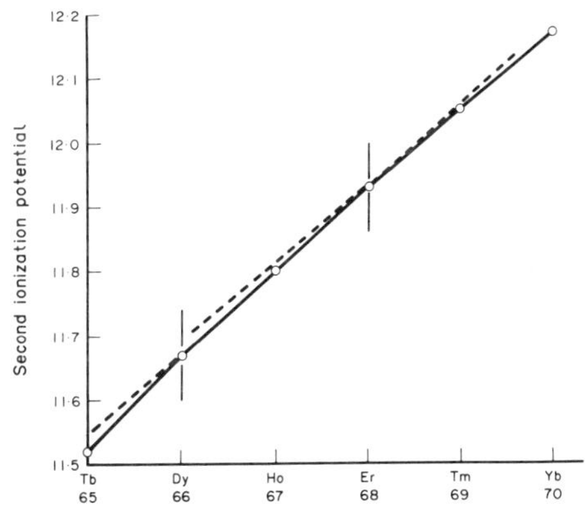

Fig. 5.6. Correlation is close and linearity appears to hold over limited ranges (three elements).

The correlations are indeed close and are confirmed by Least Squares Regression calculations on some of the Rare Earths. The calculated *IP* values are given in Table 5.2. Calculated values for y, m, x and c for the linear expression, $y = mx + c$ are also given.

TABLE 5.2

Element	At. no. (Z)	Difference (eV)	First *IP* measured (eV)	First *IP* calculated (eV)	Difference (eV)	Second *IP* measured (eV)	Second *IP* calculated (eV)
Pr	59		5.52	5.42		10.55	10.55
Nd	60	0.07	5.49	5.49	0.17	10.72	10.72
Pm	61	0.06	5.55	5.56	0.18	10.90	10.90
Sm	62	0.07	5.63	5.63	0.17	11.07	11.07
Eu	63	0.04	5.47	6.67	0.18	11.25	11.25
Tb	65		5.85	5.85		11.52	-
Dy	66	0.08	5.93	5.93	0.15	11.67	-
Ho	67	0.09	6.02	6.02	0.13	11.80	-
Er	68	0.08	6.10	6.10	0.13	11.93	-
Tm	69	0.08	6.18	6.18	0.12	12.05	-
Yb	70	0.07	6.25	6.27	0.12	12.17	-

The relevant linear equations ($y = mx + c$) are:

Pr - Eu First *IP* ; $y = 0.069x + 1.348$
Second *IP*; $y = 0.175x + 0.223$

Tb - Yb First *IP* ; $y = 0.083x + 0.455$
(excluding Yb)

Fifteen of the sixteen calculated and measured values are identical. Pm is the exception and the difference (0.01 eV) is extremely small.

Linearity holds also for the triads Dy, Ho and Er, and Er, Tm and Yb (Fig. 5.6) as the measured second *IP* values for each component pair of these triads are identical, 0.13 eV and 0.12 eV respectively. Er is common to both triads.

To the knowledge of the author linearity between *IP* and *Z* has not been reported before.

The question now arises as to the causes for linearity. The *IP*-*Z* linear relationship differs from those involving atomic and X-ray spectra. Both exhibit discrete regular changes but whereas the *IP*-*Z* "spectra" follow the expression $y = mx + c$, the "jumps" in atomic and X-ray spectra follow different and more complicated mathematical relationships. Theories of atomic and X-ray spectra have been fully investigated and the parameters are well known and account must now be taken of *IP*-*Z* linearity.

THE THIRD AND FOURTH *IPS*

Figures 5.7 and 5.8 relate the third and fourth *IP*s respectively to *Z* and a comparison of the main features of these diagrams shows clearly that they do not resemble those involving *Z* and the first and second *IP*s where the main feature is a close linearity between most *IP*s and *Z*. In this respect it should be borne in mind that each Rare Earth element contains two outermost *s* electrons and that therefore the first and second *IP*s involve these electrons. The third and fourth *IP*s involve electrons deeper in the Rare Earth ionized atoms.

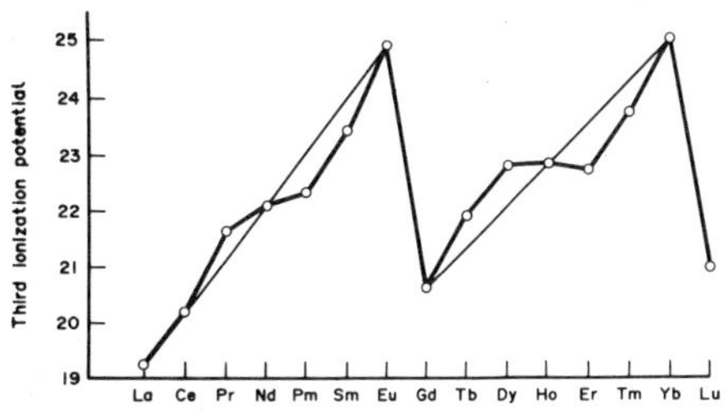

Fig. 5.7. For the third ionization potential, several well-defined features are clear including a distinct half-shell effect which is similar to that in Fig. 5.8 but differs from the half-shell discontinuities of Figs. 5.1 and 5.2.

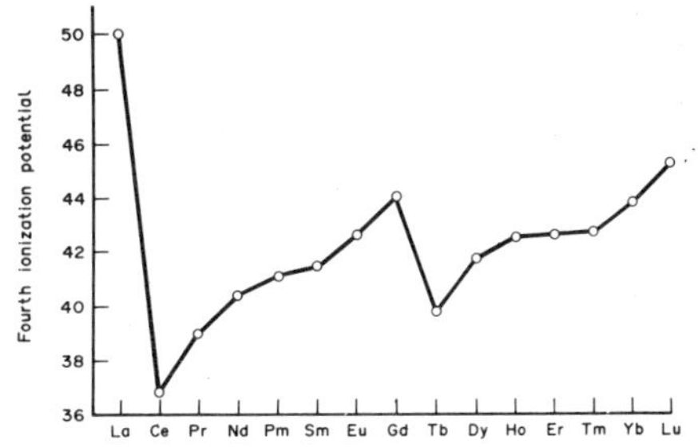

Fig. 5.8. For the fourth ionization potential, several well-defined features are clear including a distinct half-shell effect which is similar to that in Fig. 5.7 but differs from the half-shell discontinuities of Figs. 5.1 and 5.2.

It is noteworthy that the main features in Figs. 5.7 and 5.8 are similar. The two half-shell effects involving seven electrons are clear but whereas the third *IP* peaks at Eu ($Z = 63$) and falls sharply at Gd ($Z = 64$), the fourth *IP* peaks at Gd and falls sharply at Tb ($Z = 65$), a displacement of one Z.

For each half-shell of seven electrons, *IP* increases regularly with Z in quite a distinctive manner which is similar for each of the four half-shells. After an initial steep increase, the *IP*-Z relationship flattens and then rises steeply again, leading up to the peak value.

A few quantitative data may be noted:

<u>Third *IP*s</u>

First half-shell, La to Eu, increase	5.74 eV
Second half-shell, Gd to Yb, increase	4.40 "
Whole shell, La to Yb, increase	5.85 "
First half-shell decrease, Eu to Gd	4.29 "
Second half-shell decrease, Yb to Lu	4.07 "

The La-Eu join passes through Nd and the Gd-Yb join passes through Ho. The four *IP* differences which arise from the above feature are as follows:

Nd-La	2.92 eV	Ave. 2.85 eV
Eu-Nd	2.82 eV	
Ho-Gd	2.21 eV	Ave. 2.20 eV
Yb-Ho	2.19 eV	

Fourth *IP*s

First half-shell, Ce to Gd, increase	7.24 eV
Second half-shell, Tb to Lu, increase	5.40 "
Whole shell, Ce to Lu, increase	8.43 "
First half-shell decrease, Gd to Tb	4.21 "

The half-shell increases of the fourth *IP*s are greater than those of the third. The half-shell relationships involving the Rare Earth elements contrast with those involving the *d*-transition elements as the *d* half-shell differences are constant or very nearly so, 9.0 eV.

It is of interest to note that the third *IP* half-shell decreases are similar and their average, 4.18 eV, is nearly equal to the fourth *IP* half-shell decrease of 4.21 eV.

CHAPTER 6

Ionization Potential Variations in Some Isoelectronic Sequences

Lisitzin (1938) was one of the first to investigate variation of ionization potentials in some isoelectronic sequences. Several years later the same topic was discussed by Ahrens (1956). In several sequences, particularly those involving the lighter elements, the relationship between ionization potential and ionic charge is fairly simple, and may, to a good first approximation, be expressed (Lisitzin) by an equation of the type

$$I_n = aZ + bZ$$

where a and b are constants and Z ionic charge (Lisitzin). In the diagrams which follow (Figs. 6.1, 6.2 and 6.3), reproduced here after Ahrens (1956), I_n/Z is related to Z. The upper abscissal scale is for sequences involving the first s electron outside a closed shell, $Li^+ \rightarrow F^{7+}$, for example), whereas the lower scale, displaced by one unit, is for sequences involving the second s electron, $He^+ \rightarrow N^{6+}$, $Be^+ \rightarrow F^{6+}$, for example. The diagrams have been drawn this way so that careful comparison may be made of I_n/Z for pairs of s electrons; for example, Be ($I_1/1$ and $I_2/2$), B ($I_2/2$ and $I_3/3$), and so on.

Values of constants a and b for each sequence with fairly well-developed linearity are given in Table 6.1. Some values of Lisitzin are given for comparison.

It is known that for point charges ($H^+ \rightarrow F^{7+}$), $I_n = aZ^2$ where $a = R$, the Rydberg constant which is equal to the first ionization potential of hydrogen. In the 1s sequence, $He^+ \rightarrow N^{6+}$, constant a remains the same ($=R$) and $b = 10.9$ eV; that is, it is about equal to the difference between the first *IP*s of He and H (24.56 - 13.59 = 10.97 eV). This difference is hereafter indicated by R'. For 2s isoelectronic sequences $a = \sim R/4$, and b roughly $R'/5$ and $R'/2$. In the 3s sequences, a reduces to half its value for the 2s sequences and thus equals $R/8$; a reduces to 1.15 in the 4s sequences, that is, to two-thirds of 1.7 (3s sequence), and it approximately equals $R/12$. For the 5s sequences, the reduction is to 0.9 eV. For the 6s sequence starting with Au^+, the relationships between I_n/Z and Z shows much curvature (Ahrens, 1956). The variation of a in the 2s, 3s, and 4s sequences seems to follow roughly the relationship

$$a = \frac{R}{4n} = \frac{3.4}{n}$$

where n = 1, 2 and 3 for the 2s, 3s and 4s sequences respectively.

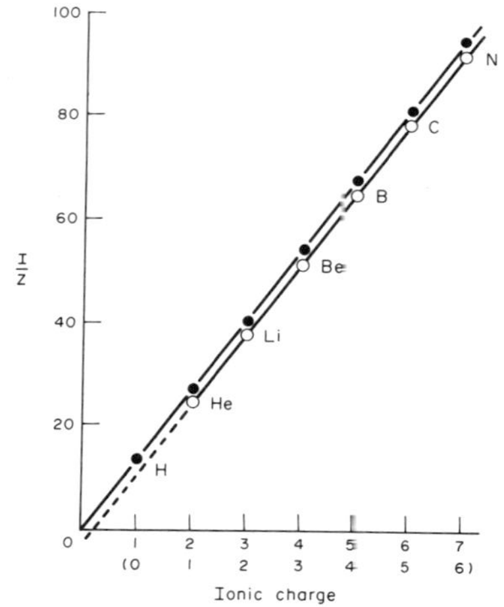

Fig. 6.1. Variation of I_n/Z with Z in the 1s isoelectronic sequences $H^+ \rightarrow N^{7+}$ (point charges - full circles) and $He^+ \rightarrow N^{6+}$, open circles. The abscissal scale for the $He^+ \rightarrow N^{+6}$ sequence is displaced by one unit, as indicated. After Ahrens (1956).

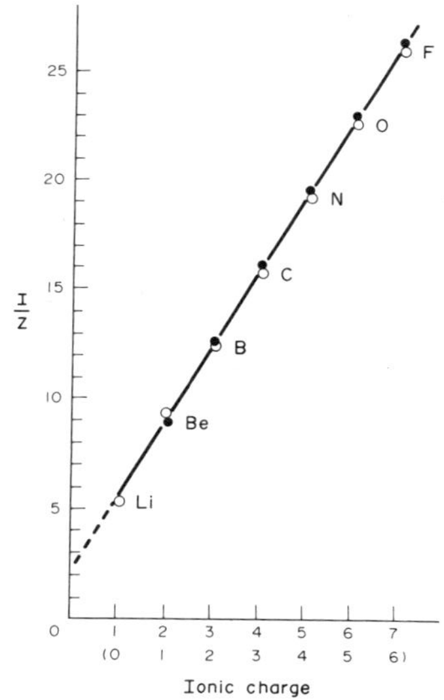

Fig. 6.2. Variation of I_n/Z with Z in the 2s isoelectronic sequences, $Li^+ \rightarrow F^{7+}$ (open circles) and $Be^+ \rightarrow F^{6+}$ (full circles). Note scale displacement as in Fig. 6.1. After Ahrens (1956).

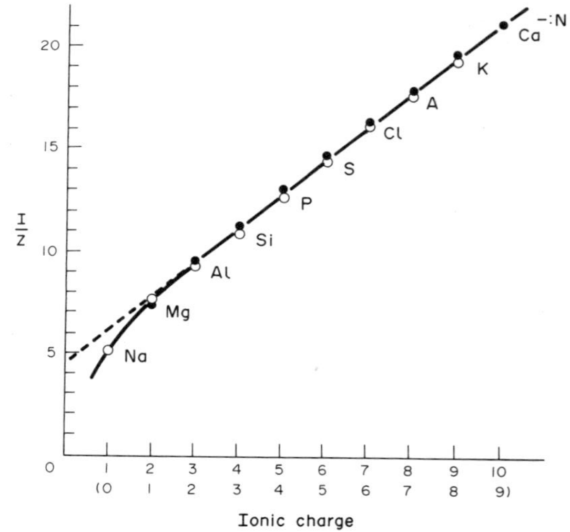

Fig. 6.3. Variation of I_n/Z with Z_a in the 3s isoelectronic sequences $Na^+ \rightarrow Ca^{10+}$ (open circles) and $Mg^+ \rightarrow Ca^{9+}$ (full circles). Note scale displacement as in Figs. 6.1 and 6.2. After Ahrens (1956).

TABLE 6.1

Values of constants *a* and *b*

Electron	Sequence	*a* Lisitzin	*a* Ahrens	*b*
1*s*	$H^+ \rightarrow N^{7+}$ (point charges)	-	13.59	-
1*s*	$He^+ \rightarrow N^{6+}$	13.54	13.59	10.0
2*s*	$Li^+ \rightarrow F^{7+}$	3.41	3.43	2.2
2*s*	$Be^+ \rightarrow F^{6+}$	3.44	3.43	5.6
3*s*	$Na^+ \rightarrow Ca^{10+}$	1.62	1.72	4.4
3*s*	$Mg^+ \rightarrow Ca^{9+}$	-	1.72	6.1
4*s*	$Cu^+ \rightarrow Kr^{8+}$	1.12	1.17	6.6
4*s*	$Zn^+ \rightarrow As^{4+}$	-	1.17	7.8
5*s*	$Ag^+ \rightarrow I^{7+}$	0.91	∿0.90	∿6.6
5*s*	$Cd \rightarrow I^{6+}$	0.94	∿0.90	∿7.6
6*s*	$Au^+ \rightarrow Po^{6+}$	0.87		
2*p*	$B^+ \rightarrow Na^{7+}$		3.5	5.4
2*p*	$C^+ \rightarrow Ne^{5+}$		3.5	7.8
2*p*	$N^+ \rightarrow Mg^{6+}$		3.5	10.0
2*p*	$O^+ \rightarrow Si^{7+}$		3.5	10.6
2*p*	$F^+ \rightarrow Al^{5+}$		3.5	13.6
2*p*	$Ne^+ \rightarrow Si^{5+}$		3.4	16.6

For sequences involving the first of a pair of *s* electrons, constant *b* appears to vary regularly in steps of 2.2 eV ($=R'/5$) to 6.6 eV ($=3R'/5$) where it appears to remain uniform for the 4*s* and 5*s* sequences.

For the 2*p* sequences (Ahrens, 1956) *a* remains uniform at ∿3.5 eV but constant *b* increases first in two steps of 2.2 - 2.4 eV and from N to O by only 0.6 eV (note cross-over) and finally in two steps of about 3 eV. Greater irregularities have been found in sequences involving 3*p* and 4*p* electrons, but as some of the *IP*s are uncertain, these sequences will not be examined.

CHAPTER 7

Oxidation States of Some Metals

Electronic structure provides the primary guide as to which oxidation (valence) states, if any, an element may exhibit. For example, the alkali metals with one *s* electron outside closed shells are monovalent and the alkaline earths with two *s* electrons outside closed shells are divalent. Atoms with closed shells, the noble gases, tend to be inert. However, some of the heaviest noble gases are not totally inert (Chapter 9) and evidently structure is not the sole controlling factor. Other difficulties may arise. Thus, Ti, Zr, Hf and Th each have the configuration d^2s^2, yet whereas Ti^{II}, Ti^{III} and Ti^{IV} are known, for Zr, Hf and particularly Th, quadrivalence only is well characterized and lower states are feeble or absent.

In their book, *Advanced Inorganic Chemistry*, Cotton and Wilkinson (1966) devote considerable space to ionization potentials and they refer to this property in their discussion on oxidation states. They write: "Although ionization potentials give some guidance concerning the relative stabilities of oxidation states, this problem is a very complex one and not amenable to facile generalization."

Much depends on the environment under consideration, crystalline solid, fused salt, melt, liquid or gas.

One main purpose in this chapter will be to ascertain to what extent *IP* differences might correlate with the existence or otherwise of stable oxidation states. The investigation can unfortunately not be thorough as the available data, particularly the higher *IP*s of the transition elements, are either approximate or lacking. Much of the discussion which follows is based on two articles by the writer (Ahrens, 1956 and 1957). He concluded that, if the difference between two successive ionization potentials, $\Delta_{n,m}$ is less than about twelve electron volts, oxidation is likely to be spontaneous. If the difference exceeds 20-22 electron volts, oxidation cannot proceed further. If $\Delta_{n,m}$ lies between these limits, oxidation states are likely to exist and their stability will in part depend on the magnitude of $\Delta_{n,m}$. The data to be considered are given at the end of the chapter and are divided into two groups, the non-transition elements and the transition elements. Cu, Ag and Au have been included with the non-transition group. The data are arranged in sequence, beginning with Group 1.

THE NON-TRANSITION METALS

We will begin with the alkali metals and the alkaline earths.

The ionization potentials of the alkali metals are extremely low (3.89 eV to 5.39 eV) and their *s* electrons are loosely bound and accordingly these elements readily react but oxidation does not proceed further than a univalent state because the second electron is firmly held as it would have to come from a closed shell. The *IP* data confirm this as their magnitudes are high. Yet, it is clear that the value of the second ionization potential decreases quite sharply from Li to Cs and the differences between the first and second ionization potential become progressively smaller and for Cs is only 21.21 eV. This value is close to the critical value (above) which suggests that Cs could almost be divalent and raises the question as to when is "a shell a shell" with regard to availability of electrons for valence purposes.

The first ionization potentials of the alkaline earths, particularly that of Be, tend to be greater than those of the alkali metals and these elements are less reactive than the alkali metals. When they do react, oxidation proceeds immediately to a divalent state but no further. The apparent reasons are that the differences between the first and second ionization potentials decrease regularly with weight. The third ionization potential of Ba is not known but extrapolation of the progressive decrease, Be to Sr, indicates a magnitude of about 25 eV, a value somewhat greater than that involving the first and second ionization potentials of Cs.

The behaviour of the Group 1 (b) and Group 2 (b) elements contrasts sharply with the alkali metals and the alkaline earths. Each contains an *s* electron outside a closed shell, but the ionization potentials indicate clearly that these *s* electrons are firmly held and that as a whole Cu, Ag and Au, and Zn, Cd and Hg are relatively inactive, particularly in the case of Au. However, if they do react oxidation may proceed beyond the univalent state in Cu, Ag and Au, and in Zn, Cd and Hg univalence is absent and divalence but not trivalence is known. As a result of this behaviour it seems desirable to critically examine the ionization potential relationships between these elements. For Cu, Ag and Au the differences between the first and second ionization potentials are relatively low (11.3 to 13.9 eV) and accordingly the second electrons in the d^{10} shells are far less tightly bound than the *p* electrons in the alkali metals and the alkaline earths (second and third ionization potentials respectively). Oxidation states greater than one are known, Cu^{II}, Ag^{II} and Au^{III}. For Zn, Cd and Hg, neither univalent nor trivalent states exist. Such behaviour accords reasonably with the ionization potential data. The differences between the first and second ionization potentials are relatively low (8-9 eV). For Zn and Cd the differences between the second and third ionization potentials are fairly high but quite close to the threshold limit referred to on page 32. For Hg, the difference is distinctly less than for Zn and Cd and the inference is that Hg could display an oxidation state greater than 2, but no such state is known to exist.

Hg appears to be univalent as in calomel but this is not due to the species Hg^{I}, but rather to Hg_2^{2+}. The existence of this anomaly might be associated with the fact that the first ionization potential of Hg is conspicuously high (Ahrens, 1956, and in prep.).

Al, Ga, In and Tl (s^2p^1)

The observed facts are: (i) that trivalence is characteristic for each element, (ii) neither divalence nor states greater than three exist, and (iii) that whereas Al^I is unknown and a spurious Ga^I seemingly exists, univalence in In and particularly Tl is well characterized.

Observation (i) accords with ionization potential data as the differences between the third and fourth ionization potentials (28 to 31 eV) are sufficiently great to prevent oxidation beyond trivalence. Observation (ii) likewise accords with ionization potential data as the differences between second and third ionization potentials are low (9-10 eV). With regard to (iii), $\Delta_{1,2}$ lies within the limits listed, stated on p. , and Ga^{II} might be anticipated. "Divalence" of Ga is known as in $GaCl_2$ but is due to a possible "mixture" of Ga^I and Ga^{III} and not the presence of Ga^{II}.

Si, Ge, Sn and Pb (s^2p^2)

The observed facts are as follows. A well-characterized state of IV but not more, absence of states I and III and the existence of divalence. The absence of oxidation states greater than IV agrees with the fact that $\Delta_{4,5}$ is relatively high, and the absence of states I and III is also in agreement with ionization potential data; for each element $\Delta_{1,2}$ and $\Delta_{3,4}$ are low. The presence of divalence is also in agreement with ionization potential data; for each element, $\Delta_{2,3}$ has an intermediate value. However, ionization potentials do not support the fact that the stability of divalence increases with weight from feeble in Si to well-characterized valence in Sn and particularly Pb. For "neighbours" Si and Ge, *IP* data do support the fact that Ge^{II} is more stable than Si^{II}.

P, As, Sb and Bi (s^2p^3)

Phosphorus is a non-metal but has nevertheless been included for the sake of comparison. As for the Group 4 and Group 3 elements, the observed behaviour correlates reasonably well with the ionization potential data. States greater than 5 do not exist due to a high value of $\Delta_{5,6}$ and absence of univalence, divalence and quadrivalence may be ascribed to relatively low ionization potential differences. $\Delta_{3,4}$ is intermediate between the differences referred to above and trivalence, particularly of the heavy metals, is characteristic.

Sc, Y and La (d^1s^2)

The characteristic oxidation state for each of these elements is three and three only, for two reasons: (i) $\Delta_{3,4}$ is high and $\Delta_{2,3}$ is low for each element. A possible oxidation state of two for Sc might be anticipated because the $\Delta_{2,3}$ value of 12 eV is close to the lower threshold limit of 10-12 eV, but there appears to be no evidence of divalence. For Y and La, $\Delta_{2,3}$ is low, < 10 eV.

Ti, Zr, Hf and Th (d^2s^2).

These Group 4 elements make an interesting comparison. Each element exhibits its Group valence of four and for Th, no other states are evidently known. Ti^{II} and Ti^{III} are distinctly stable states, but divalence and trivalence in Zr and Hf is either feeble or absent. Such behaviour accords reasonably well with ionization potential differences.

CONCLUDING COMMENTS

Taken as a whole *IP* differences may serve as a semi-quantitative guide for indicating the presence or absence of oxidation states of the metallic elements, and, in addition, may serve as a semi-quantitative guide of stability. Several exceptions exist and in part these may be due to what has been described as a "weight effect" (Ahrens, 1957).

TABLE 7.1

IP data including *IP* differences

(a) Non-transition elements

	I_1	I_2	$\Delta_{1,2}$		I_1	I_2	$\Delta_{1,2}$
Li	5.39	75.64	70.25				
Na	5.14	47.29	42.15				
K	4.34	31.63	27.28	Cu	7.73	20.29	12.57
Rb	4.18	27.28	23.10	Ag	7.58	21.49	13.91
Cs	3.89	25.10	21.21	Au	9.23	20.50	11.28

	I_1	I_2	I_3	$\Delta_{2,3}$		I_1	I_2	I_3	$\Delta_{2,3}$
Be	9.32	18.21	153.89	135.68					
Mg	7.65	15.04	80.14	65.11					
Ca	6.11	11.87	50.91	39.04	Zn	9.39	17.96	39.72	21.76
Sr	5.70	11.03	43.(600)	32.57	Cd	8.99	16.91	37.48	20.57
Ba	5.21	10.00	-	(25)	Hg	10.44	18.76	34.20	15.44

	I_1	I_2	I_3	I_4		$\Delta_{1,2}$	$\Delta_{2,3}$	$\Delta_{3,4}$
Al	5.99	18.83	28.45	120.0	Al	12.8	9.62	165
Ga	6.00	20.51	30.71	64	Ga	14.5	10.20	33.3
In	5.79	18.87	28.03	54	In	13.1	9.16	26
Tl	6.11	20.43	29.83		Tl	14.3	9.40	-

	I_1	I_2	I_3	I_4	I_5		$\Delta_{1,2}$	$\Delta_{2,3}$	$\Delta_{3,4}$	$\Delta_{4,5}$
Si	8.15	16.35	33.49	45.14	166.8	Si	8.20	17.14	11.7	121.7
Ge	7.90	15.93	34.22	45.71	93.5	Ge	8.03	18.29	11.5	47.8
Sn	7.34	14.63	30.50	40.73	72.28	Sn	7.29	15.87	10.2	31.6
Pb	7.42	15.03	31.94	42.32	68.8	Pb	7.61	16.91	10.4	26.5

TABLE 7.1 (*cont.*)

	I_1	I_2	I_3	I_4	I_5	I_6
P	10.49	19.73	30.18	51.37	65.02	220.4
As	9.81	18.63	28.35	50.13	62.63	127.6
Sb	8.64	16.53	25.30	44.2	56	108
Bi	7.29	18.69	25.56	45.3	56.0	88.3

	$\Delta_{1,2}$	$\Delta_{2,3}$	$\Delta_{3,4}$	$\Delta_{4,5}$	$\Delta_{5,6}$
P	9.24	10.45	21.2	13.65	43.0
As	8.82	9.72	21.8	12.50	49.9
Sb	7.89	8.77	18.9	11.8	52
Bi	11.40	6.87	19.7	10.7	32.3

(b) Transition elements

	I_1	I_2	I_3	I_4	$\Delta_{2,3}$	$\Delta_{3,4}$
Sc	6.54	12.80	24.76	73.47	11.96	48.71
Y	6.38	12.24	20.52	(61.8)	8.28	(41.28)
La	5.58	11.06	19.18	-	8.12	-

	I_1	I_2	I_3	I_4	I_5
Ti	6.82	13.58	27.49	43.27	99.22
Zr	6.84	13.13	22.99	34.34	81.50
Hf	7.0	14.9	23.33	33.3	-
Th	6.95	11.5	20.0	28.8	-

	$\Delta_{1,2}$	$\Delta_{2,3}$	$\Delta_{3,4}$	$\Delta_{4,5}$
Ti	6.76	13.91	15.78	55.95
Zr	6.29	9.85	11.35	47.16
Hf	7.9	8.4	10	-
Th	4.55	8.5	8.8	-

CHAPTER 8

Screening (Shielding)

The concept of screening (shielding) of the positive charge on the nucleus by electrons has been discussed by many physicists and chemists. Some use the term shielding efficiency. There is little doubt that shielding is a vital atomic property for several reasons: it is one major parameter which determines the magnitude of the effective nuclear charge and of cationic forces (polarizing power), Chapter 10, and is helpful when considering chemical inertness. Although there is broad agreement as to what is understood by screening, difficulties arise when attempting to define this quantity in precise and quantitative terms.

Some workers assign specific screening numbers to different electron types, Coulson (1952) for example. A few excerpts from his book follow.

"To determine the shielding constant, electrons are divided into the following groups, each having a different shielding constant: 1*s*; 2*s*, 2*p*; 3*s*, 3*p*; 3*d*; 4*s*, 4*p*; 4*d*; 4*f*; 5*s*, 5*p*; etc. Thus the *s* and *p* of a given shell are grouped together, but the *d* and *f* are separated. The shells are considered to be arranged from inside out in the order named." After a formulation of his rules, Coulson states: "They do, however, become somewhat unreliable as soon as orbitals with a total quantum number 4 is reached, and must then be used with caution."

In a detailed and thorough work, Lakatos, Bohus and Medgyesi (1959) used screening numbers to determine effective field strength and effective nuclear charge. They state, "The electron groups with the inert gas configuration ($1s^2$ and s^2p^6) have the most perfect screening of all the groups". They state further that the groups p^3, d^5 and f^7 also screen relatively well (each is a half-shell) and that, "The positive charge on the nucleus is, in general, very imperfectly screened by the 18-electron ($s^2p^6d^{10}$) and 32-electron ($s^2p^6d^{10}f^{14}$) groups with non-inert gas configuration and by electron groups with unclosed orbits". In their paper Lakatos, Bohus and Medgyesi refer to the works of Slater, Angus and Pauling.

In his work on ionic radii, Pauling used X-ray term values for the purpose of estimating screening.

All the above publications on screening are important but they have not been discussed in detail as the *IP* has not been taken into account in their calculations.

Finkelnburg (1950) was one of the first to utilize the *IP* to determine some function of screening. His calculations extend from hydrogen to element 36, Kr. An estimate of screening was not made but values of screening constants and screening increments are given. Values for the effective nuclear charge are also listed (see his Table 9). Finkelnburg used the first *IP* only and I have difficulty in assessing the value of his determinations. A similar comment was made by Lakatos *et al.*

Shielding efficiency of cations has been considered by Ahrens (1954). He states that, "The main object of this communication is to suggest a quantity which may be used for *comparing* the shielding efficiency (S_{eff}) of all cations . . ." *IP*, charge and ionic radius are taken into consideration.

The 8- and 18-electron cations, Na^+ and Cu^+, will be considered first. Both carry unit charge and have similar ionic radii; yet, whereas only 5.1 eV is required to remove an electron from an atom of sodium in sodium vapour, 7.7 eV is required to ionize copper to Cu^+. As the $4s$ electron in copper is more firmly held than the $3s$ electron in sodium, the 8-electron shell screens more efficiently than the 18-electron shell.

In the above example, singly charged cations were considered but for more highly charged cations, a higher *IP* will be used - the second for doubly charged cations, the third for trebly charged cations, and so on: thus, comparisons are always made of the energy required to remove one electron - the last one - to give the particular ionic state in question.

As a second example, we may take the pair Zr^{4+} and Hf^{4+}, each with a charge of four and a radius of about 0.79 Å; their fourth *IP*s are 34.34 eV and 33.3 eV, respectively. Hence, as it requires more energy to remove the $4d$ electron from outside the 36-electron krypton core than to remove the $5d$ electron from outside the 68-electron hafnium core, giving in each case a cation of equal charge and radius, shielding in Zr^{4+} is apparently slightly less efficient than in Hf^{4+}.

In the above examples, radius and charge are the same for the constituents of each pair and nothing can be concluded about how the shielding efficiency in Na^+ compares with that in Zr^{4+}, where both radius and charge differ. Consider first, the relationship between r and *IP* when charge is kept constant. Ahrens (1954) observed that the relationship

$$I \propto 1/\sqrt{r}$$

is quite closely followed by the 8-electron groups, Na^+, K^+, Rb^+, Cs^+, and Mg^{2+}, Ca^{2+}, Sr^{2+} and Ba^{2+}. Neither the 2-electron cations, Li^+ and Be^{2+}, nor the 18-electron cations, Cu^+, Ag^+, Au^+ and Zn^{2+}, Cd^{2+} and Hg^{2+}, fit the above general relationship. Attention was also directed to smooth inverse regularities which exist between r and I_n in such sequences as $Li^+ \rightarrow F^{7+}$; $Na^+ \rightarrow Cl^{7+}$; $K^+ \rightarrow Mn^{7+}$; $Cu^+ \rightarrow Br^{7+}$; $Ag^+ \rightarrow I^{7+}$, etc. Combination of such regularities with the above inverse-square relationship yields

$$\sqrt{r}I = KZ^{1 \cdot 27},$$

where Z is the ionic charge. With *IP* in eV and r in Å, K is found to be 5.00 ± 0.01.

It has been suggested (Ahrens, 1954) that the quantity $5Z^{1.27}/\sqrt{rI}$ may be used as a measure of the shielding efficiency of cations, and those cations which obey the above equation and therefore have S_{eff} = 1.00 (Table 8.1) will be defined as having *normal* shielding efficiency.

Table 8.2 gives values of S_{eff} for some cations arranged according to the Periodic Table. Ahrens relates S_{eff} to atomic number in consecutive sequences; his Fig. 1.

TABLE 8.1

Calculated values of K and S_{eff}

	$Z^{1.27}$	r(A.)	I_n (eV)	K	S_{eff}
Na^+	1	0.95	5.12	4.99	1.00
Mg^{2+}	2.41	0.65	15.0	5.01	1.00
Al^{3+}	4.03	0.51	28.5	5.05	0.99
Si^{4+}	5.82	0.42	45.2	5.05	0.99
P^{5+}	7.72	0.35	65.0	4.98	1.00
S^{6+}	9.73	0.30	88.1	4.97	1.005
Cl^{7+}	11.83	0.27	114.3	5.00	1.00
K^+	1	1.33	4.34	5.01	1.00
Ca^{2+}	2.41	1.01	11.9	4.96	1.01
Rb^+	1	1.46	4.17	5.03	0.99
Sr^{2+}	2.41	1.18	11.0	4.95	1.01
Cs^+	1	1.66	3.89	5.01	1.00
Ba^{2+}	2.41	1.34	10.0	(4.80)*	(1.04)
Ra^{2+}	2.41	1.43	10.1	4.98	1.00
				Av. 5.00	1.00

*Omitted from calculation of the average.

TABLE 8.2

Shielding efficiency of cations

Li^+	Be^{2+}	B^{3+}		C^{4+}	N^{5+}	O^{6+}	F^{7+}
1.12	1.12	1.10		1.13	1.10	1.12	1.13
Na^+	Mg^{2+}	Al^{3+}		Si^{4+}	P^{5+}	Se^{+6}	Cl^{7+}
1.00	1.00	0.99		0.99	1.00	1.005	1.00
K^+	Ca^{2+}	Sc^{3+}		Ti^{4+}	V^{5+}	Cr^{6+}	Mn^{7+}
1.00	1.01	0.90		0.82	0.77	0.74	0.71
Cu^+	Zn^{2+}	Ga^{3+}		Ge^{4+}	As^{5+}	Se^{6+}	Br^{7+}
0.68	0.80	0.86		0.90	0.96	0.97	0.98
Rb^+	Sr^{2+}	Y^{3+}		Zr^{4+}	Nb^{5+}	Mo^{6+}	Tc^{7+}
0.99	1.01	1.03		0.96	0.93	0.92	0.85
Ag^+	Cd^{2+}	In^{3+}		Sn^{4+}	Sb^{5+}	Te^{6+}	I^{7+}
0.59	0.725	0.80		0.86	0.83	0.92	0.93
Cs^+	Ba^{2+}	La^{3+}	Lu^{3+}	Hf^{4+}	Ta^{5+}	W^{6+}	Re^{7+}
1.00	1.04	0.99	1.14	1.06	1.04	1.02	1.00
Au^+	Hg^{2+}	Tl^{3+}		Pb^{4+}	Bi^{5+}	Po^{6+}	At^{7+}
0.46	0.61	0.70		0.75	0.81	(0.86)	(0.90)
	Ra^{2+}			Th^{4+}			
	1.00			0.98			

S_{eff} either remains uniform ($Li^+ \rightarrow F^{7+}$, and $Na^+ \rightarrow Cl^{7+}$) or varies regularly toward or away from a value of unity. Entry of $3d$ electrons in particular, and $4d$ and $5d$ to a lesser extent, cause a lowering in the shielding efficiency as defined; the effect increases with charge. The reverse developments take place after completion of the 18-electron shells and in the sequences $Cu^+ \rightarrow Br^{7+}$, $Ag^+ \rightarrow I^{7+}$ and $Au^+ \rightarrow At^{7+}$, S_{eff} increases with charge and tends to approach unity in each case.

According to Ahrens (1954) the effect of entry of $4f$ electrons is apparently unique. S_{eff} increases from about unity to 1.14, a maximum value which is apparently very similar in magnitude to that of the 2-electron cations.

For his 1954 paper Ahrens used the *IP* data given in the 1950 Tabellen of Landolt-Börnstein.

CHAPTER 9

Relatively Inert Elements

This is a two-part chapter in which the importance of *IP*s are emphasized for the purpose of gaining an insight into the inactivity of the noble gases and gold. The noble gases are considered in the first part and gold in the second.

SECTION 1. The noble gases

Reluctance to react with other elements is characteristic of the "inert" gases He, Ne, Ar, Kr, Xe and Rn and it is for this reason that they have been styled the Noble Gases. The concensus view of "the day" (late 19th century to mid-20th century) was that the s^2 duplet (He) and the s^2p^6 octet of Ne, Ar, Kr, Xe and Rn were so stable that so-called duet and octet rules were formulated and were regarded as inviolate. Such an attitude inhibited research into the forming of noble gas compounds although L. Pauling had in 1933 predicted their formation (Heslop and Jones, 1976). Transient compounds and unstable hydrates and clathrates have been known for some time, but not chemical compounds as usually understood.

During a course of inorganic chemistry lectures given to second- and third-year students at the University of Cape Town, the author drew attention (class notes of 1957-1960) to the facts that,

(i) the first *IP*s of the noble gases were relatively high (relative to the alkali metals, for example).

(ii) *IP*s decreased regularly from He to Xe.

(iii) It was pointed out that because the *IP*s of the heaviest noble gases were not "very high", they might combine with such a highly active element as fluorine.

First *IP*s of the noble gases are given in Table 9.1.

TABLE 9.1

First ionization potentials of the noble gases (eV)

	Electron configuration	Ionization potential	
He	2 $1s^2$	24.58	
Ne	10 $2s^2p^6$	21.56	
Ar	18 $3s^2p^6$	15.76	Apparent threshold
Kr	36 $4s^2p^6$	14.00	level
Xe	54 $5s^2p^6$	12.13	
Rn	86 $6s^2p^6$	10.75	

The first noble gas compounds were reported in 1962 by Bartlett and by Claasen, Selig and Mann (Heslop and Jones, 1976) and it is not surprising that such compounds have been confined to the heaviest noble gases and that they are mostly fluorides and, to some lesser extent, chlorine and oxygen compounds. Some complex compounds are also known. Because of its low half-life, Rn compounds are difficult to produce and identify. Xe forms the largest and most stable compounds. A few Kr compounds are known and these are relatively unstable.

Several of the *IP*s of Xe are not accurately known and, likewise, some of those of Kr are uncertain. It may be useful, however, to consider those of Ar as they are accurately known and may be instructive in an indirect way. These are given in Table 9.2, together with differences between successive ionization potentials, $I_{n,m}$.

TABLE 9.2

Ionization potentials of argon (eV)

	I_1		I_2		I_3		I_4		I_5		I_6		I_7		I_8		I_9
	15.76		27.63		40.73		59.81		75.02		91.01		124.32		143.46		422.4
$I_{n,m}$		11.87		13.1		19.07		15.2		15.99		33.3		19.15		279	

The only correlations which are evident are "steps" between I_3 and I_4 which corresponds to a sub-shell effect involving *p* and *s* electrons and I_6 and I_7, a half-shell effect and I_8 and I_9, a whole-shell effect.

SECTION 2. Gold

The unusual chemistry of gold, as manifested by its inertness, its tendency to form complexes, and its high heat of hydration, has been recognized by many and is referred to in several texts. Chemical inertness is the most striking feature of this coinage metal, a feature which is considered here. When doing so, some attention has been drawn to trends involving Hg, Tl, Pb and Bi elements which follow Au in the Periodic Table (Ahrens, in prep.).

Although attention has indeed frequently been drawn to the exceptional nobility of gold, less attention has apparently been given to why this should be and why in particular Au is chemically less reactive than Cu and Ag. This feature contrasts strongly with the typical activity trends of Li, Na,

K, Rb and Cs and of Be, Mg, Ca, Sr and Ba. Clearly one or more atomic properties must account for such an anomalous behaviour and the *IP* appears to serve as one such useful property.

When considering the geochemistry of gold, Ahrens (1964) drew attention to its strong siderophile tendency (see Chapter 16) and its high first *IP*. It was pointed out that because the first ionization potential was relatively high for a metal, its outermost s^1 electron was firmly bound and this apparently accounted for its reluctance to react chemically and its strong tendency to exist therefore in an uncombined state in a geological environment. A similar approach will be considered here, but in greater detail and depth and is similar to that used in Section 1. Although the approaches used in Sections 1 and 2 are similar, precise *comparisons* will not be undertaken because elements with vastly different electronic structures are involved.

First *IP*s of Cu, Ag and Au are given in Table 9.3, together with those of the alkali metals. Both groups of elements contain *s* electrons outside closed shells. Ionic radii are also given.

TABLE 9.3

First ionization potentials and ionic radii of the alkali metals and the coinage metals

	I_1 (eV)	radius (Å)		I_1 (eV)	radius (Å)
Li	5.34	0.68			
Na	5.18	0.97			
K	4.34	1.34	Cu	7.7	0.9
Rb	4.18	1.44	Ag	7.5	1.3
Cs	3.89	1.67	Au	9.23	1.4

Several features are clear.

(i) The first *IP*s of the alkali metals fall regularly with atomic weight;

(ii) radii vary regularly and inversely with *IP*;

(iii) the first *IP*s of the coinage metals are all distinctly greater than those of the alkali metals;

(iv) ionic radii* of Cu^+, Ag^+ and Au^+ decrease regularly with weight and;

(v) whereas the *IP* of Ag is less than that of Cu, in accordance with the typical trends in the Periodic Table, as shown by the alkali metals in Table 9.3, that of Au is elevated relative to Cu and Ag; and

(vi) such an increase which is marked, is counter to the typical trends and may be regarded therefore as an anomaly.

The relatively high first *IP*s of the coinage metals may be taken to be the reason for their relative inactivity, and their siderophile tendency (Chapter 16). Whereas these generalizations are in themselves important, the Cu, Ag-Au discontinuity is particularly noteworthy. Had the normal trend persisted, the *IP* of gold would have been less than silver and its chemistry

---

*The radii of these ions might not be accurate but they serve as reasonable magnitudes.

would probably have been approximately midway between that of silver and thallous thallium. Gold would not be a noble element, would not be siderophile, would probably have been associated with sulphur in sulphide minerals and could not be used for jewellery. Had gold followed the typical trends in the Periodic Table, the consequences would have been vast, economically, politically and socially.

CHAPTER 10

Cationic Forces (Polarizing Power)

Several terms which are more or less synonymous have been suggested for cationic forces; for example, polarizing power, effective force of attraction of a cation, ionic potential, binding power of a cation and anion affinity. Electronegativity is a related property, but, by usual definition, does not refer specifically to cationic forces; we come to this property again shortly. For the purpose of further discussion the term polarizing power will be chosen because it refers specifically to polarization, which for our purpose of bond discussion is apposite. It should be borne in mind, however, that when used in another context, namely that involved in metal binding with organic ligands (see Chapter 12), the term *binding power* is more appropriate.

It was recognized in the middle 1920s (Goldschmidt, 1926) that two properties, ionic radius and cationic charge (Z), were of particular significance for estimating the magnitude of the positive force of attraction associated with a cation. It was suggested that the ratio Z/r^2 could serve as an appropriate function and a similar function, Z/r, known as the ionic potential, was recommended by Cartledge (1928; 1930). A somewhat more complex function involving charge and radius is the so-called EK-value suggested by Fersman (1935) in his studies on the geochemical distribution of the elements. The EK-value is defined as

$$EK = K \frac{Z^2}{2r} \cdot$$

Szádeczky-Kardoss (1954, 1955) suggested a function (the Verbindungspotential) which embraced radius and charge. This function, defined as

$$\frac{a \dfrac{Z1}{r^1} + b \dfrac{Z2}{r^2} + c \dfrac{Z3}{r^3}}{a \quad b \quad c} ,$$

refers to all constituent ions in a crystal.

Functions involving the ratio of charge to radius work reasonably well as a measure of cationic force (and polarizing power) for certain purposes, and provided only ions with similar electronic structures are considered. For example, Ramberg (1953) has pointed out that the ionic potential is a useful measure of cation field strength for the noble-gas-type cations. However, if different electronic structure types are involved serious anomalies may arise

(Ahrens, 1953; Goldschmidt, 1954). Thus, for example, the ionic potentials of each component in the pairs

$$Na^+–Cu^+;\ K^+–Ag^+\ (Au^+);\ Mg^{2+}–Zn^{2+};\ Ca^{2+}–Cd^{2+}\ \text{and}\ Sr^{2+}–Hg^{2+}$$

are similar, whereas the polarizing power of the 18-electron cation, the second in each pair, is far greater than that of the noble gas ion. In order to explain this "anomaly". a third property must be considered, namely that of variable screening (shielding). Different electron arrangements screen the positive charge on the nucleus with different efficiency and accordingly influence the resultant forces acting on the periphery of the atom or ion and hence the polarizing power of a cation.

Ahrens (1953) suggested the quantity

$$I_n/Z$$

as a measure of cationic force and in his book, *Geochemistry*, Goldschmidt (1954) also recommended this ratio in a detailed section entitled, "The ionization potential as a measure of the polarizing force of ions". I_n is the ionization potential of the last (*ultimate*) electron. Lakatos *et al.* (1958) use the term *effective electric field strength* (F^+) and define this quantity according to the relationship

$$F^+ = Z_{eff} \cdot e/r^2$$

where r is the radius of the "isolated species" (ion or atom) and should not therefore be regarded as a radius in the solid state and Z_{eff} the effective nuclear charge. It is stated further: "The effective field strength is the actual electric attractive force exerted on a unit negative charge located at r distance from the nucleus, or according to another definition, F^+ is proportional to the actual amount of charge per unit surface of atom or ion with radius r."

As a measure of the magnitude of cationic forces, Ahrens (1953; 1961) and Ahrens and Morris (1956) preferred to use the nth ionization potential as such.

An indication of the sensitivity dependence of I_n on radius and charge may be obtained by inspecting the data in Table 10.1.

TABLE 10.1

Some selected ionization potentials:
(a) charge constant and (b) radius approximately constant

	r(Å)	I_2		r(Å)	Charge	I_n
	(a)			(b)		
Be^{2+}	0.35	18.2	Na^+	0.97	1	5.14
Mg^{2+}	0.65	15.03	Ca^{2+}	1.01	2	11.87
Ca^{2+}	1.01	11.87	Y^{3+}	0.92	3	20.5
Sr^{2+}	1.18	11.03	Th^{4+}	1.02	4	∿29.7
Ba^{2+}	1.34	10.00				

In (a) atoms of a similar kind are considered and it may be seen that for an approximately two-fold increase in radius (Mg^{2+}, 0.65 Å to Ba^{2+}, 1.34 Å) the second ionization potential drops by one-third. It is clear from Table 10.1 that the magnitude of I_n is very sensitively dependent on charge and increases by more than a factor of 2 if charge is doubled. (Y and Th are transition-type ions and should for strict comparison not be included in

Table 10.1.) Because, however, non-transition-type ions of high charge and having a radius magnitude of 1 Å do not exist, and as the shielding efficiency in Y and Th appears to be quite high and similar to that obtaining in 8-electron atoms, Y and Th serve reasonably well for our purpose.

The *n*th ionization potential will be used in the discussions in this book which involve polarizing power. When used in this way it is desirable to keep charge constant and radius as uniform as possible.

THE EFFECT OF ELECTRONIC STRUCTURE ON POLARIZING POWER

The importance of electronic structure on screening was discussed in Chapter 7. It follows from the manner in which the ionization potential has been used when defining screening that the magnitude of the polarizing power will also in part be determined by electronic structure. The relationships between 8- and 18-electron cations are compared below. In these comparisons, charge will be kept constant and radius approximately uniform. The radii should be regarded as magnitudes. The *n*th ionization potential is given in brackets after each cation (Table 10.2).

TABLE 10.2

Radius	-	1 Å	Na^+	(5.1)	Cu^+	(7.7)
"	-	1.25-1.45 Å	K^+	(4.4)	Ag^+	(7.9)
			Rb^+	(4.2)	Au^+	(9.3)
Radius	-	0.7 Å	Mg^{2+}	(15.03)	Zn^{2+}	(17.9)
"	-	1 Å	Ca^{2+}	(11.9)	Cd^{2+}	(16.9)
"	-	1.15 Å	Sr^{2+}	(11.03)	Hg^{2+}	(18.2)
Radius	-	0.5-0.6 Å	Al^{3+}	(28.4)	Ga^{3+}	(30.7)
Radius	-	0.4-0.5 Å	Si^{4+}	(45.1)	Ge^{4+}	(45.7)

It is clear that for a given charge and a similar size, the polarizing power of an 18-electron cation is greater than that of an 8-electron cation: no exceptions exist. It is clear also that differences on a relative basis vary enormously, from a small difference (Si and Ge) to one that is extremely large, Rb and Au. For this latter pair, the ratio of the first ionization potentials (Au/Rb) is more than 2.

The second *IP*s of the 3*d* elements with incomplete shells and half-shells, for a given size magnitude of 0.7 Å, are Fe (16.18), Co (17.06), Ni (18.17) and Cu (20.29). These values are all greater than Mg; some are less and some greater than Zn.

NOTES ON ELECTRONEGATIVITY

Numerous workers have attempted to define electronegativity since the concept was introduced a few decades ago by L. Pauling. Ahrens (1953) has criticized the concept in certain respects and uses *IP*s where possible in its stead. It is not the purpose here to debate the merits and demerits of this concept but to draw attention to some comments which relate electronegativity and *IP*.

Fyfe (1954) has discussed the importance of the transition between extreme ionic and covalent character of a bond in mineralogy and writes, "It should be noted that essentially there is little difference between the approach to bond transition from consideration of electronegativities to the approach via polarizability of Fajans and others which Ahrens uses. Consideration of charge and radius effects give an order of bond transition similar to that derived from electronegativities." Ringwood (1955) made similar comments: "Ahrens (1953) and Goldschmidt (1954) have suggested the use of ionization potential as a relative measure of the ionic or covalent character of the chemical bond. This usage which is based on the polarization model of transitional chemical bonding (Fajans) also gives satisfactory results and could be used to replace the electronegativity criterion if so desired." Finally, Mitra (1961) states that, "ionization potentials may in any case serve as an approximate measure of electronegativity".

CHAPTER 11

Heats of Hydration

Several criteria determine the magnitude of the heat of hydration of a cation. These include charge, radius, *IP* and crystal field stabilization. As our primary emphasis here is on the *IP* it is desirable to determine the significance of the other factors but in so doing it must be borne in mind that the magnitude of the *IP* itself depends on charge and size. In other words, it is not possible to ascertain quantitatively to what extent *IP* alone contributes to the total heat of hydration unless radius and charge are kept as constant as possible. Let us consider first some pairs of ions of given charge and similar radius and which do not exhibit crystal field stabilization; that is, those cations which have complete shells or complete half-shells. Some examples are given below. As it is customary to express heats of hydration (-HH) in kcals this unit will be used in place of electron volts. The data are those given by Basolo and Pearson (1959).

Singly charged cations

(i) Radius magnitude 1 Å

	IP	-HH
Na^+	118.5	100
Cu^+	178	139

(ii) Radius magnitude, 1.3-1.5 Å

	IP	-HH
K^+	100.1	79
Rb^+	96.3	75
Au^+	213	154

Doubly charged cations

(i) Radius magnitude 1 Å

	Total *IP*	-HH
Ca^{2+}	414	382
Cd^{2+}	597	437

(ii) Radius magnitude 0.7 Å

Mg^{2+}	523	464
Zn^{2+}	631	492
Cu^{2+}	646	507

Cu^{2+} with a configuration of d^9 has been included for the sake of comparison as this cation exhibits crystal field stabilization.

The examples above indicate clearly that for a given charge and size, heat of hydration increases with ionization potential.

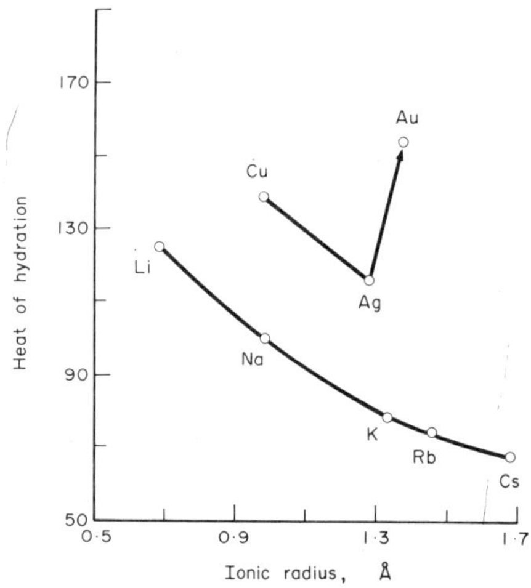

Fig. 11.1. In this relationship between heat of hydration and ionic radius several features are clear; see details in text.

Figure 11.1 relates ionic radius to heats of hydration for the singly charged cations Li^+ (124.3), Na^+ (118.5), K^+ (100.1), Rb^+ (96.3), Cs^+ (89.8), Cu^+ (178), Ag^+ (175) and Au^+ (215). *IP*s are given in brackets. For the alkali metals heat of hydration decreases regularly with increase of radius and decrease of *IP*. The coinage metals define a "field" well above that of alkali metal trends. The Au anomaly is of considerable interest. Radius in the sequence Cu^+, Ag^+ and Au^+ increases, but instead of a decrease in the first *IP* of Au, there is a distinct increase and with it an increase in the heat of hydration of Au. The Group 2 elements display features similar to those of the Group 1 elements. As above, total *IP*s are given in brackets. Mg^{2+} (523), Ca^{2+} (414), Sr^{2+} (385), Ba^{2+} (351), Zn^{2+} (631), Cd^{2+} (597) and Hg^{2+} (673). The anomaly at Hg, similar to that of Au, is clearly evident.

Au and Hg follow the entry of the *4f* electrons and have anomalously high *IP*s and many consequences arise from this development including the marked increase of ionization potential and heat of hydration.

Although heats of hydration tend to correlate with total *IP*s the relationship is not simple even if *d*-transition elements with unfilled shells or half-shells are excluded. Two examples will be considered; the singly charged cations Li^+, Na^+, K^+, Rb^+, Cr^+, Cu^+, Ag^+, Au^+ and Tl^+; and the doubly charged cations Mg^{2+}, Ca^{2+}, Sr^{2+}, Ba^{2+}, Sn^{2+}, Cd^{2+}, Hg^{2+} and Mn^{2+}.

Figure 11.2 relates *IP* to -HH for the above singly charged cations and Fig. 11.3 refers to the doubly charged cations. Several features are apparent.

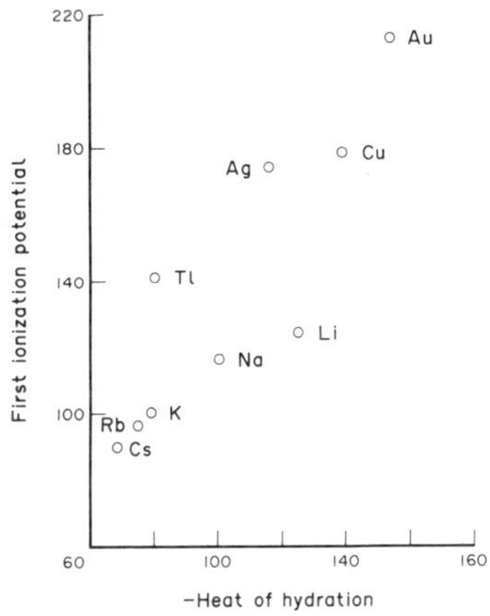

Fig. 11.2. The relationship between ionization potential and heat of hydration for singly-charged cations.

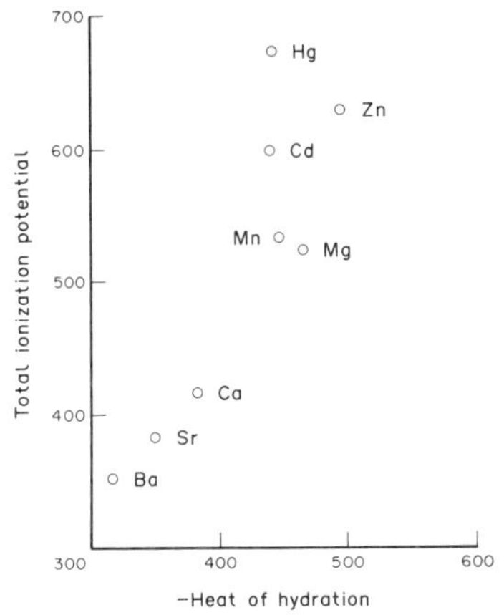

Fig. 11.3. The relationship between total ionization potential and heat of hydration for doubly-charged cations.

In Fig. 11.2 two sets of relationship are apparent, one involving the alkali metals and the other the heavy metals. In both sets, ionization potential correlates with heats of hydration but the heavy metal trend is displaced. For these elements the increases in heats of hydration are less than that of the ionization potentials. This relationship becomes apparent when the respective ratios, *IP*/-HH are compared, as follows:

Singly charged cations

(i) Li^+ (0.994) Na^+ (1.185) K^+ (1.267) Rb^+ (1.284) Cs^+ (1.320)

(ii) Cu^+ (1.280) Ag^+ (1.508) Au^+ (1.383) and Tl^+ (1.762).

Doubly charged cations

(i) Mg^{2+} (1.127) Ca^{2+} (1.084) Sr^{2+} (1.10) Ba^{2+} (1.11)

(ii) Zn^{2+} (1.282) Cd^{2+} (1.366) Hg^{2+} (1.526) Mn^{2+} (1.195) Pb^{2+} (1.443), Sn^{2+} (1.353).

IP to heat ratios are shown as a histogram in Fig. 11.4. Most ratios fall between 1.0 to 1.4, the modal value is approximately 1.25. Five ions have ratios greater than 1.4 and these arranged in sequence of increasing ratio are: Pb^{2+} (1.443), Ag^{+} (1.508), Hg^{2+} (1.526), Ag^{2+} (1.657) and Tl^{+} (1.762). Each element with a high ratio is a heavy metal.

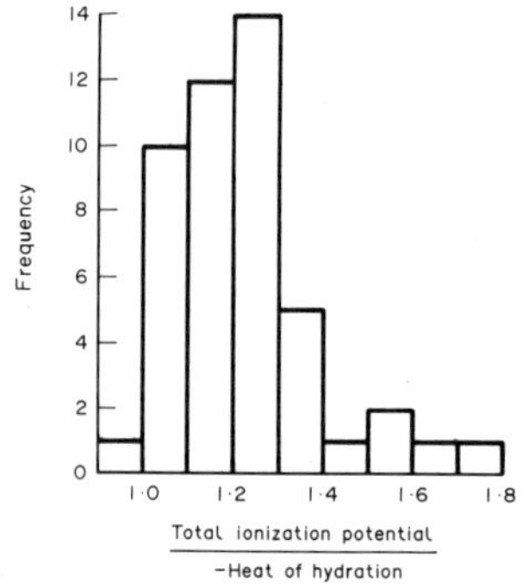

Fig. 11.4. Frequency distribution diagram involving total ionization potential and heat of hydration. The spread of values is quite considerable and the high (modal) point is approximately 1.25.

CHAPTER 12

The Stability of Metal-Organic Complexes

PROPERTIES OF METALS, OTHER THAN IONIZATION POTENTIAL, WHICH MAY INFLUENCE METAL-COMPLEX STABILITY

For given conditions (temperature, pH, etc.) the tendency to form metal-organic complexes and the stability of such complexes depends on properties both of the metal atom and of the organic ligand; for example, its structure, including ability to form ring (chelate) complexes, polarizability and types of available bonding atoms (O, S, N). As, however, the emphasis here is on the use of the ionization potential of the metal, the discussion will be devoted mainly to properties of metals, though the nature of the organic ligand must naturally be borne in mind.

Several properties of a metal atom influence metal-ligand stability and consideration should be given first to the importance of properties other than the ionization potential in order that the significance of this specific property may be more clearly revealed. As in other studies the approach will be to keep all other properties as constant as possible when attempting to examine the specific importance of one given property of the metal atom.

The oxidation state or ionic charge is one highly significant property. In order to illustrate clearly the importance of charge, it is desirable for this property to vary widely; cationic radius should remain as uniform as possible. A search through lists of available cationic radii shows that for ions of approximately constant radius, there is only one group in which charge varies by as much as four; namely, Na^+ (0.97 Å), Ca^{2+} (1.01 Å), Rare Earth cations (RE^{3+}) (0.9-1.15 Å) and Th^{4+} (1.02 Å). Only a few organic ligands form stable complexes with all of these cations. The well-known analytical reagent ethylenediamine-*NNN'N'*-tetra-acetic acid (EDTA) is one, and as complete data on this reagent are available, it may serve as an example to provide a general indication of the effect of charge variation on metal-complex stability. Stability data are given in Table 12.1, and Fig. 12.1 relates the logarithm of the stability constant (K) to ionic charge. The high sensitivity dependence of the stability of the complexes on charge is clearly demonstrated by the data of Table 12.1 and the graphical representation in Fig. 12.1.

TABLE 12.1

Log K for EDTA complexes of M^{+}, M^{2+}, M^{3+}, and M^{4+}, cations with a radius magnitude of ∿1.0 Å

Cation	Log K
Na^{+}	1.7
Ca^{2+}	10.7
RE^{3+}	15.5-19.8 (Av. 17.5)
Th^{4+}	23.2

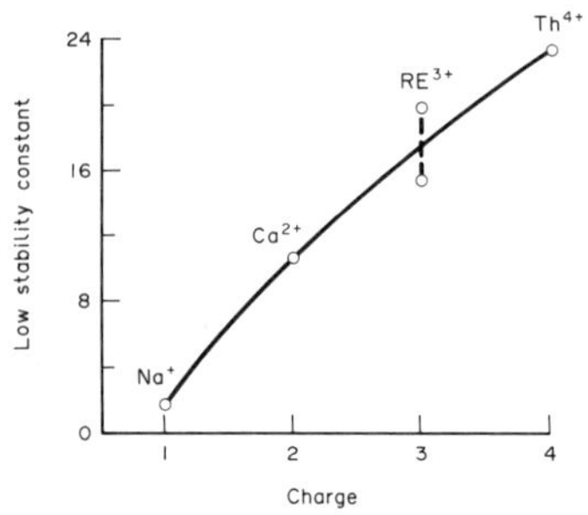

Fig. 12.1. The relationship between log stability constant and charge indicates clearly the sensitivity of stability on cationic charge. Note the location of the Rare Earth elements.

The curve of best fit passes about mid-way between the limits of the Rare Earth range; the mid-point corresponds to a cation radius magnitude of ∿1 Å. The variation of RE^{3+}-EDTA stability with radius is itself an illustration of the dependence of complex stability on cation radius, an aspect which is considered below.

Because stability depends so sensitively on charge, authors frequently group metals according to charge when attempting to set up a sequence for metal-complex stabilities. For example, Basolo and Pearson (1958) list three groups of sequences according to the charge on the metal. For divalent metals they propose the following sequence: Pt > Pd > Hg > (UO_2) > Be > Cu > Ni > Co > Pb > Zn > Cd > Fe > Mn > Mg* > Ca > Sr > Ba > (*Mg has been added).

Metals with different electronic configurations and cations with different radii (range, from 0.35 Å for Be^{2+} to 1.34 Å for Ba^{2+}) are included in this sequence, and for our purpose of gaining as much insight as possible about the significance of a specific property of a metal we should regroup the above elements according to size magnitude in the first place. Many of the cations of the above metals fall into two size magnitudes, 1.0 Å (±0.15 Å) and 0.7 Å (±0.1 Å), and the stability relationships, according to the sequence listed above, in each size magnitude, are *usually* as follows:

(i) Hg^{2+} (1.12 Å) > (Pb^{2+}) (1.20 Å) > Cd^{2+} (0.97 Å) > Ca^{2+} (1.01 Å) > Sr^{2+} (1.15 Å)

(ii) Pt^{2+} (0.70 Å) > Pd^{2+} (0.80 Å) > Cu^{2+} (0.80 Å) > Ni^{2+} (0.69 Å) > Co^{2+} (0.72 Å) > Sn^{2+} (0.69 Å) > Fe^{2+} (0.74 Å) > Mn^{2+} (0.80 Å) > Mg^{2+} (0.65 Å)

The importance of size can be illustrated by considering the alkaline earths. Radii vary from 0.35 Å (Be^{2+}) to 1.34 Å (Ba^{2+}). They form stable complexes with several organic ligands including the oxalic acid ion. Stability data are given in Table 12.2.

TABLE 12.2

Stability constants of alkaline earth-oxalate complexes arranged in order of increasing radius of cation

	Radius (Å)	Log stability constant
Be^{2+}	0.35	∿4.0
Mg^{2+}	0.65	3.43
Ca^{2+}	1.01	3.00
Sr^{2+}	1.18	2.54
Ba^{2+}	1.35	2.33

It is clear that for the oxalic acid ion, stability falls steadily with increase in radius and though this relationship is not perfectly general, it applies to several organic ligands; for the three cations, Ca^{2+}, Sr^{2+} and Ba^{2+}, the above size-stability relationship holds fairly generally.

THE IMPORTANCE OF IONIZATION POTENTIAL

Irving and Williams (1953) were among the first to use the *IP* for the purpose of studying the stabilities of metal-organic complexes. Several others, including Ahrens (1966), have used the *IP* for the same purpose. Ahrens examined in some detail the stabilities of metal-amino acid complexes and for the medium-sized cations (Mg^{2+}, Mn^{2+}, Fe^{2+}, Co^{2+}, Zn^{2+}, Ni^{2+} and Cu^{2+}) reported a close correlation between *IP* and stability. Six amino acids were studied and the average values of the log stability constant are related to *IP* in Fig. 12.2.

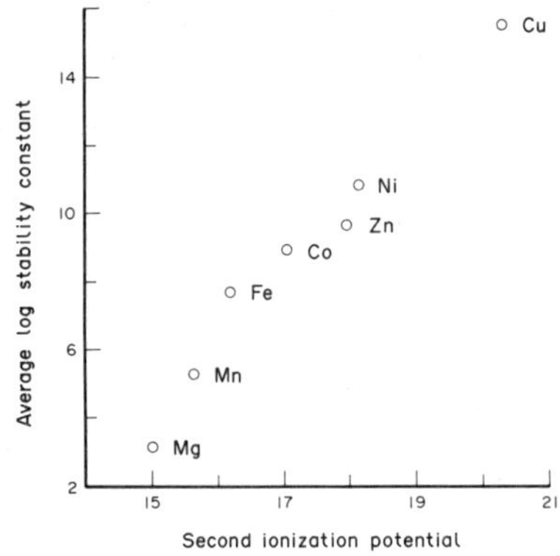

Fig. 12.2. The relationship between the average log stability constant of six amino acids and the second ionization potential.

Though the regularity is not fully quantitative, it is nevertheless good - perhaps surprisingly so - when it is borne in mind that atoms with distinctly different structures are involved; 8- and 18-electrons full-shell ions (Mg^{2+} and Zn^{2+}, respectively), a 3*d* half-shell ion (Mn^{2+}) and several 3*d* ions with incomplete shells (Fe^{2+}, Co^{2+}, Ni^{2+} and Cu^{2+}). Compared with the ionization potentials, electronegativities do not correlate so well with complex stabilities (Ahrens, 1966).

One generalization applies with very few exceptions both to the amino acids and to other organic ligands; namely, that for singly, doubly and possibly more highly charged cations, the stability of the complex with the 18-electron cation is greater than that with an 8-electron cation of like charge and similar radius. Thus

Charge^{1+}	$Cu^+ \gg Na^+$	($r \sim 1$ Å)
	Ag^+ and $Au^+ \gg K^+$ and Rb^+	($r \sim 1.25$–1.45 Å)
Charge^{2+}	$Zn^{2+} > Mg^{2+}$	($r \sim 0.7$ Å)
	$Cd^{2+} > Ca^{2+}$	($r \sim 1.0$ Å)
	$Hg^{2+} \gg Sr^{2+}$	($r \sim 1.15$ Å)
Charge^{3+}	$Ga^{3+} > Al^{3+}$	($r \sim 0.50$ Å)
	$In^{3+} > Sc^{3+}$	($r \sim 0.80$ Å)
	$Tl^{3+} > RE^{3+}s$	($r \sim 1.0$ Å)

These relationships accord with ionization potential differences. Relatively, ionization potential differences and stability differences are greatest for singly charged cations and become progressively less as charge increases.

It is of considerable interest to note that in a fairly general way, the heaviest elements of the groups Cu-Ag-Au and Zn-Cd-Hg, namely Au and Hg, show particularly strong tendencies to form complexes with a wide variety of ligands. This is apparently due to the fact that in each group, the heaviest element has a conspicuously high ionization potential.

Though the stabilities of the amino acid complexes with the medium sized divalent elements (Mg^{2+} . . . Cu^{2+}) correlate quite well with the respective ionization potentials, the relationships are sometimes less satisfactory when other organic ligands are considered.

Sometimes, for example, the position of neighbouring Co and Zn may be such that the increase in stability of the Zn-ligand complex is only slightly greater than that of the Co-ligand complex, despite the fact that the second ionization potential of Zn (17.96 eV) is quite distinctly greater than that of Co (17.05 eV). Sometimes, in fact, the Co-Zn relationship may even be reversed: an example of each of the above is briefly considered below.

Figure 3 of Ahrens (1966) relates I_2 (medium-sized cations) to the log-stability constants for ethylenediamine-*NN'*-dipropionic acid and ethylenediamine-*NNN'N'*-tetrapropionic acid. Though stability increases with the second ionization potential for each of seven metals, the increase at Zn, relative to Co, is very slight indeed.

Dimethylglyoxime ($C_4H_8O_2N_2$), the well-known analytical reagent, may serve as an example to illustrate a Co-Zn reversal. Log-stability constants ($\log \beta_2$) are 13.9 for Zn and 18.9 for Co; stability increases for Ni (21.7) and reaches a maximum at Cu (23.3).

The Co-Zn "effect" is evidently due to ligand-field stabilization. Such apparent stabilization is confined to the transition metals but will not be considered further here because we are concerned primarily with the importance of ionization potentials. The importance of ligand-field effects is discussed in numerous papers and books including the one by Basolo and Pearson (1958).

CHAPTER 13

Ionization Potentials and the Chemical Binding and Structure of Simple Inorganic Crystalline Compounds — I

CHEMICAL BINDING

INTRODUCTION

In the interpretation of the problem of transitions between electrovalency and covalency, the concept of resonance between ionic and covalent structures and the principle of the electronegativity of elements are usually employed.[1,2]

In the present paper the nature of bonding of metal-nonmetal crystals is considered from an ionic point of view. In this treatment it is assumed that, in the formation of crystals of inorganic salts, a hypothetical stage occurs, which involves interaction of the constituents as ions, whether the resulting compound is predominantly ionic or covalent. According to this approach, which is somewhat similar to that of Fajans,[3,4] it appears that only when cations exert a sufficiently large electrostatic field strength on their neighbours will a crystal lattice with bonds with covalent character be produced.

Ahrens[5] has suggested the use of ionization potentials I (first for univalent cations, second for divalent cations, etc.) as a measure of the relative electrostatic field strengths of cations of the same formal charge in the formation of crystalline compounds.* One considerable advantage in using ionization-potential data is that accurate values are available for most of the cations with which we are concerned. Since the ionization potential is a measure of the energy required to remove a single electron from a free atom

*Goldschmidt[6] in *Geochemistry* (Oxford, 1954) has suggested the use of values of (I/formal charge) as a measure of the polarizing force exerted by cations in crystals.

(1) L. Pauling, *The Nature of the Chemical Bond*, Cornell (1940).
(2) C. A. Coulson, *Proc. Roy. Soc.* A207, 63 (1951).
(3) K. Fajans, *Naturwiss.* 11, 165 (1923).
(4) K. Fajans and G. Joos, *Z. Physik.* 23, 1 (1924).
(5) L. H. Ahrens, *Geochim. et Cosmochim. Acta* 3, 1 (1953).
(6) V. M. Goldschmidt, *Geochemistry*, Oxford (1954).

or ion, it is also a measure of the energy lost by the recombination of the resulting ion and the electron, and hence may be taken as indicative of the attraction of the ion on the electron last removed. Although the ionization potential refers to free ions, it seems reasonable, when considering the formation of crystals from different cations of the same formal charge and (crystal) radius with a given anionic constituent, to employ the ionization potentials of the cations as an approximate comparative measure of their power of attraction for the most loosely bound electrons of the neighbouring anions. The greater the power of the cations for drawing the most loosely bound electrons of neighbouring anions to themselves, the greater the deviations from extreme ionic character that may be expected in the resultant crystal. According to this rather oversimplified picture, one might expect to observe transitions from essentially ionic towards essentially covalent character in crystalline compounds, regarded as initially formed from cations of the same formal charge and radius with a common anion, as the I values of the cations become greater.

With regard to the comparison of the above ionic approach to bonding in crystals with the electronegativity concept of Pauling, it should be pointed out that his electronegativity values refer to stable equilibrium conditions, whereas the method which considers ionic interaction may be regarded as referring to a transitory stage governing the properties of the resulting crystal.*

IONIZATION POTENTIALS AND TRANSITIONS BETWEEN ELECTROVALENCY AND COVALENCY IN SIMPLE INORGANIC CRYSTALLINE COMPOUNDS

In this section, some of the changes in bond character of crystalline compounds, regarded as initially formed from cations of the same formal charge and similar ionic (crystal) radii, are pointed out. Ideal comparisons are only possible when both the formal charge and size of the cations are equal, e.g. Zr^{4+} (r = 0.79 Å) and Hf^{4+} (r = 0.78 Å). In the present case, however, grouping has been carried out according to range of size, although radius differences must be taken into account in a detailed discussion.

In most cases it is not easy to ascribe to the bonding of inorganic crystalline compounds a definite amount of ionic or covalent character, and statements based on thermochemical electronegativity data such as ". . . Beryllium bonds have the following amounts of ionic character: Be-F, 79%, BE-O, 63%; Be-Cl, 44%; Be-Br, 35%; Be-I, 22% . . ."(1) can only be taken as giving a *rough* guide (see Pritchard and Skinner(10)). For this reason, the transitions between electrovalency and covalency will be dealt with in a qualitative or semiquantitative way in this paper, and various properties which are widely regarded as giving indications of the ionic or covalent character of the compounds will be mentioned. Relations between ionization potentials and the different crystal structures are discussed in a second paper.

---

*In this connection, it is interesting to remember that Mulliken(7,8) has pointed out that the average of the *first* ionization potential and the electron affinity should measure the electronegativity of an atom. See Baughan(9) and Pritchard and Skinner(10) for a further discussion of this relationship.

(7) R. S. Mulliken, *J. Chem. Phys.* 2, 782 (1934).
(8) R. S. Mulliken, *J. Chem. Phys.* 3, 573 (1935).
(9) E. C. Baughan, *Quart. Reviews* 7, 103 (1953).
(10) H. O. Pritchard and H. A. Skinner, *Chem. Reviews* 55, 745 (1955).

Crystalline compounds of univalent metals

Table 13.1 shows univalent cations arranged in order of increasing first ionization potential of the corresponding atoms, together with their crystal radii.* For some purposes of comparison, Na^+ and Cu^+ should be omitted, and a more satisfactory sequence is obtained by considering only the larger cations Rb^+, K^+, Tl^+, Ag^+, and Au^+.

TABLE 13.1

Univalent cations, ionization potentials and radii

	Cs^+	Rb^+	K^+	Na^+	Tl^+	Ag^+	Cu^+	Au^+
I (volts)	3.893	4.176	4.339	5.138	6.106	7.574	7.724	9.223
r (Å)	1.65	1.44	1.33	0.95	1.47	1.26	0.91	1.37

A study of the properties of crystalline salts regarded as formed from the cations listed in Table 13.1 shows that a definite increase in covalent character occurs with increase in I. Thus when the I value of the metallic element is less than *c.* 5.5 volts, not only the fluorides and chlorides but also the oxides, other halides, and sulphides are dominantly ionic anhydrous compounds. Thallium (I = 6.106 V) in its univalent compounds begins to show definite deviation from truly ionic character, and shows chemical resemblances to its nearest large neighbours in Table 13.1 - rubidium, potassium, and silver. Departure from extreme ionic behaviour is readily illustrated, for example, by the low melting-point (300°C) of the oxide Tl_2O. The thallous halides, however, are predominantly ionic in the crystalline state; this is suggested for TlCl by microwave data.[14] Silver, cuprous copper, and aurous gold compounds are more covalent in character, the only simple crystalline compound which is essentially ionic being silver fluoride.

The differences Δ between the values of the lattice energies of crystals calculated from experimental thermodynamic data (U) and theoretical values (U_0) calculated assuming ionic crystals with a van der Waals potential, have been used to give an indication of deviations from extreme ionic character. In Table 13.2 data are given on the experimental and theoretical lattice energies of univalent halides, according to the ionization potentials of the electropositive constituents. The remarkable agreement between values of U and U_0 for the alkali halides provides strong support for the thesis that the forces operative in these crystals are those which underlie the theoretical calculations of U_0. The Δ values for TlCl, TlBr, and TlI suggest some departure from strict ionic character in these compounds. As the ionization potential of the cationic constituent becomes greater, as in the case of the argentous and cuprous halides, $U-U_0$ becomes more strongly positive, implying increased departure from strict heteropolarity. The lattice-energy data thus clearly indicate the trend for transitions from extreme ionic character to occur with increase in I.

*Ionization potential values employed throughout this paper are from Finkelnburg and Humbach[11] and are in electron volts. Ionic radii are from Ahrens[12] as modified, with additions, after Pauling[1] and Ahrens[13].

(11) W. Finkelnburg and W. Humbach, *Naturwiss.* 42, 35 (1955).

(12) L. H. Ahrens, Unpublished data (1955).

(13) L. H. Ahrens, *Geochim. et Cosmochim. Acta* 2, 155 (1952).

(14) A. P. Altshuller, *J. Chem. Phys.* 22, 1136 (1954).

TABLE 13.2

Ionization potentials and the difference between lattice energies (kcal/mole) derived from experimental thermodynamic data and theoretical values for ionic crystals

	Substance	I (electropositive constituent)	U	U_0	$\Delta = U-U_0$
		V			
← I increases	CsF	3.893	173	176	-3
	RbF	4.176	185	183	+2
	KF	4.339	195	193	+2
	NaF	5.138	217	215	+2
	TlF	6.106	198		
	AgF	7.574	231	219	+12
	CuF	7.724			
	AuF	9.223			
← I increases	CsCl	3.893	155	153	+2
	RbCl	4.176	162	162	0
	KCl	4.339	171	168	+3
	NaCl	5.138	185	184	+1
	TlCl	6.106	176	167	+9
	AgCl	7.574	219	203	+16
	CuCl	7.724	234	216	+18
	AuCl	9.223	249		
← I increases	CsBr	3.893	151	150	+1
	RbBr	4.176	157	156	+1
	KBr	4.339	165	161	+4
	NaBr	5.138	178	176	+2
	TlBr	6.106	173	164	+9
	AgBr	7.574	217	197	+20
	CuBr	7.724	232	208	+24
	AuBr	9.223	249		
← I increases	CsI	3.893	143	143	0
	RbI	4.176	149	148	+1
	KI	4.339	156	152	+4
	NaI	5.138	167	164	+3
	TlI	6.106	168	159	+9
	AgI	7.574	214	190	+24
	CuI	7.724	229	199	+30
	AuI	9.223	251		

U values have a probable error of ±2-3 kcal, and have been calculated using the Born-Haber cycle from what is thought to be the best available thermodynamic data. It should be noted that these calculations employ values for the electron affinities of halogens from Pritchard[15], which are based on about ten independent experimental determinations. See Morris[16] for further details on the values for the alkali halides.

U_0 values for the alkali halides have been taken from Huggins,[17] for silver and thallous halides from Mayer,[18] and for cuprous halides from Mayer and Levy.[19] The values have a possible error of ±2-4 kcal even if the assumptions on which their calculation is based hold exactly.

(15) H. O. Pritchard, *Chem. Reviews* 52, 529 (1953).
(16) D. F. C. Morris, *Acta Cryst.* 9, 197 (1956).
(17) M. L. Huggins, *J. Chem. Phys.* 5, 143 (1937).
(18) J. E. Mayer, *J. Chem. Phys.* 1, 327 (1933).
(19) J. E. Mayer and R. B. Levy, *J. Chem. Phys.* 1, 647 (1933).

Crystalline compounds of divalent metals

Table 13.3(a) and (b) shows divalent cations arranged in order of increasing second ionization potential of the corresponding atoms. The divalent ions are separated into two size groups. The division is arbitrary, but is based in part on the fact that the 6:3 rutile structure is common to the fluorides of the medium-sized group, whereas the 8:4 fluorite structure prevails in fluorides of the large-sized cations.

TABLE 13.3

Divalent cations: ionization potentials and crystal radii

(a) Medium-sized (r 0.6–0.9 Å) divalent cations

	Mg^{2+}	Mn^{2+}	Fe^{2+}	Co^{2+}	Zn^{2+}	Ni^{2+}	Pt^{2+}	Pd^{2+}	Cu^{2+}	Ag^{2+}
I (volts)	15.03	15.64	16.18	17.05	17.96	18.15	18.54	19.42	20.29	21.48
r (Å)	0.64	0.80	0.74	0.72	0.69	0.69	0.80	0.80	0.70	0.89

I increases ———

(b) Large-sized (r 0.95–1.35 Å) divalent cations

	Ba^{2+}	Sr^{2+}	Ca^{2+}	Pb^{2+}	Cd^{2+}	Hg^{2+}
I (volts)	10.00	11.026	11.87	15.03	16.904	18.751
r (Å)	1.35	1.18	1.02	1.20	0.97	1.10

I increases ———

Among the halides regarded as formed from the medium-sized divalent cations, those of magnesium appear to be predominantly ionic, although the bromide and iodide exhibit solubility in organic solvents. Interatomic-distance measurements and magnetic data* also suggest that all the halides of divalent manganese, iron, cobalt, and nickel have predominantly ionic character. In the case of the halides of divalent palladium (I = 19.2 V), the crystalline

*It should be pointed out here that the interpretation of the magnetic criterion for bond type should be treated with some reserve. For example, Pauling[20] seems to have abandoned his original interpretation, and wrote in 1948:
"We conclude accordingly that the magnetic criterion distinguishes, not between essentially covalent bonds and essentially ionic bonds, but between strong covalent bonds, using good hybrid bond orbitals and with the possibility of unsynchronised ionic-covalent bonds, using poor bond orbitals, and the necessity for synchronisation of the covalent phases of the bonds."

While the precise meaning of Pauling's statement appears difficult to understand, some idea of the confusion on the subject may be gained by comparing the above statement with the account of Martell and Calvin,[21] who point out some of the limitations of the magnetic method, but then conclude that, when the method may be applied, it indicates whether the bonds are "essentially covalent" or "essentially ionic". (See Selwood.[22])

(20) I. Pauling, *J. Chem. Soc.*, 1461 (1948).
(21) A. E. Martell and M. Calvin, *Chemistry of the Metal Chelates*, Prentice-Hall, New York (1952).
(22) P. W. Selwood, *Ann. Rev. Phys. Chem.* 4, 445 (1953).

substances $PdCl^2$ and PdI^2 are diamagnetic and can reasonably be regarded as possessing square essentially covalent bonds. From Hund's rules, the Pd^{2+} ion should have two unpaired *4d* electrons and should therefore have a magnetic moment of 2.83 BM calculated on the spin-only formula. If the bonding in the palladium halides was essentially ionic, therefore, they would be expected to be paramagnetic. In the case of the cupric and argentic halides, the magnetic criterion cannot be readily applied, since for both ionic and covalent configurations one unpaired electron may be expected. However, inter-atomic distance measurements and the crystal structure suggest that the bonding in cupric chloride crystals, for example, is essentially covalent.

Among the oxides regarded as formed from the medium-sized doubly-charged cations, magnesium oxide appears to be dominantly ionic, although X-ray studies have revealed that the ions in this compound, in contrast to those in sodium chloride, for example, are not entirely discrete.(23,24) Ionic character is also evident in the oxides of ferrous iron (I = 16.18 V) and cobaltous cobalt (I = 17.05 V). Zinc oxide shows considerable deviation from ionic character, and the electron-density diagram shows that the electrons are not arranged symmetrically around the zinc and oxygen atoms as if they were in the ionic form, but there is an increased density between the zinc atom and the nearest oxygen atom of the irregular tetrahedral arrangement.(25,26,27) Nickelous oxide also shows deviations from ionic character and has a distorted structure below 200°C, while the oxides of divalent platinum, palladium, and copper are dominantly nonionic.

Among the monosulphides of the medium-sized divalent cations, essentially ionic character is shown up to manganese (I = 15.64 V), and this appears to be the case with the disulphides also. Thus, for example, although MnS_2 and FeS^2 both have the pyrite structure, the available magnetic data for MnS_2, interpreted by the Weiss-Curie equation, leads to the value of 6.1 BM for the magnetic moment of manganese, in close agreement with the calculated spin moment of 5.92 BM for Mn^{2+}, whereas FeS_2 is only slightly paramagnetic and cannot correspond to an ionic structure, since the calculated spin moment of Fe^{2+} is 4.90 BM.

Similar transitions from essentially ionic character to predominantly covalent character with increase in I are shown in the compounds of the large-sized divalent cations of Table 13.3(b).

In many cases the increase in I is accompanied by a deviation in the colours of the crystals from the colours of their components, which, according to Pitzer and Hildebrand,(28) may be taken as an indication of covalency. An illustration of this effect is given in Table 13.4 by the colours of the oxides, sulphides, and iodides of the large-sized cations of Table 13.3(b).

---

(23) R. Brill, H. G. Grimm, C. Hermann and C. Peters, *Ann. Physik* 34, 393 (1939).
(24) C. Peters, *Z. Electrochem.* 46, 436 (1940).
(25) H. J. Yearian, *Phys. Rev.* 48, 631 (1935).
(26) K. Lark-Horowitz and C. H. Ehrhardt, *Phys. Rev.* 55, 605 (1939).
(27) H. M. James and V. A. Johnson, *Phys. Rev.* 56, 119 (1939).
(28) K. S. Pitzer and J. H. Hildebrand, *J. Amer. Chem. Soc.* 63, 2472 (1941).

TABLE 13.4

Colour and covalency among crystalline compounds of large-sized divalent cations

	Compound	Colour of crystals	Bond distance, Å: Measured	Bond distance, Å: Ionic radii sum (corrected where necessary)	
I increases ↓	BaO	colourless	2.75	2.75	ionic character decreases ↓
	SrO	colourless	2.54	2.53	
	CaO	colourless	2.40	2.39	
	PbO	red, yellow	2.30	*c.* 2.6	
	CdO	brown	2.34	2.37	
	HgO	red, yellow			
I increases ↓	BaS	colourless	3.18	3.19	ionic character decreases ↓
	SrS	colourless	3.00	2.97	
	CaS	colourless	2.83	2.83	
	PbS	black	2.96	3.04	
	CdS	yellow	2.51, 2.53	2.61	
	HgS	red, black	2.52, 2.91, 3.25 (red),	2.94	
			2.53 (black)	2.73	
I increases ↓	BaI^2	colourless			ionic character decreases ↓
	SrI^2	colourless			
	CaI^2	colourless	3.12	3.15	
	PbI^2	yellow	3.13		
	CdI^2	brownish	2.99	3.13	
	HgI^2	red, yellow	2.78 (red)	3.10	

The trend from typically ionic character towards covalency parallel to increasing ionization potential, which has been illustrated for compounds of univalent and divalent elements, is also observed in the crystalline compounds of elements forming cations of higher formal charge. When cations of a given oxidation number and range of size are considered, the levels of ionization potential at which considerable deviations from ionic character occur in crystals appear to depend largely upon the electron affinity and polarizability of the anionic constituent. Thus the fact that higher *I* values are necessary to produce deviations from ionic character in the case of fluorides than with oxides, sulphides, and other halides may be correlated with the high electron affinity of fluorine and the low polarizability of the fluoride ion.

CHAPTER 14

Ionization Potentials and the Chemical Binding and Structure of Simple Inorganic Crystalline Compounds — II

CRYSTAL STRUCTURE

INTRODUCTION

The concept of a crystal being built up by packing of spheres, with the arrangement of ions in the crystal determined by radius ratio [r (cation)/r (anion)], has proved suitable for describing the structures of many simple inorganic compounds in which the bonding is predominantly ionic. In the case of crystalline compounds where the bonding is predominantly covalent in character, an explanation of the structures is commonly sought in terms of orbital hybridization.[1,2,3]

Although resonance-hybridization theory frequently seems able to predict structures successfully, exceptions are quite common. In addition, it is appropriate to point out that hybridization is only a convenient language, and to quote from Coulson[3] (p. 219): "We must not, however, allow ourselves to believe that it (hybridization) represents any real phenomenon any more than resonance between different structures such as the covalent and ionic ones of a polar bond may be called a phenomenon."

In the present paper an attempt is made to explore another possible approach to the problem of non-ionic structures (e.g. those not in accordance with the radius-ratio rules) of simple metal-non-metal compounds. It is not our intention, however, to attempt to formulate a new theory or hypothesis, but merely to indicate possibly significant relationships which might ultimately be helpful for formulating a generally satisfactory theory. As in Part I, it is assumed that the formation of simple inorganic crystalline compounds proceeds via an ionic stage, whether the resulting compounds are structurally ionic or not. Accordingly, the structure as well as the nature of the chemical bonding of the equilibrium product may be expected to depend, to some significant extent at least, on the properties of the interacting cation

---

(1) L. Pauling, *The Nature of the Chemical Bond*, Cornell (1940).
(2) A. F. Wells, *Structural Inorganic Chemistry*, Oxford (1950).
(3) C. A. Coulson, *Valence*, Oxford (1952).

and anion. It will be assumed in accordance with the suggestions of Fajans, Weyl, Goldschmidt, Ahrens[(4)] and others, that polarizability, particularly that of the anion, as it is usually more deformable than the cation, is an important property. For the cation the magnitude of its effective electrostatic field strength is considered the most important property, but other properties such as cation polarizability and a possible non-spherical charge distribution, even in the free cation, may also be significant. As before, the *n*th ionization potential (I) will be used as a measure for comparing the effective field strengths of cations of given formal charge (n) and (crystal) radius (r). It is easy, of course, to find several elements yielding cations of the same formal charge, but unfortunately not with precisely the same radius: consequently, elements will be grouped according to oxidation number and size magnitude, but the possibility that slight size variation may be important will not be overlooked.

CRYSTALLINE COMPOUNDS OF DIVALENT METALS

The compounds of elements forming medium-sized cations (r = 0.64–0·90 Å) are considered first.

The crystal energies of simple crystalline compounds, regarded as formed via such cations, are shown in Table 14.1, together with the second ionization potentials of the metallic constituents. The fact that the crystal energies fairly regularly increase with I immediately suggests the possible importance of I values in relation to crystal structure.* Indeed, it is found that the crystal structures of the compounds do vary fairly systematically according to the value of I.

TABLE 14.1

Divalent cations: ionization potentials and crystal energies (U) medium-sized (r 0.6–0.9 Å) divalent cations

	Mg^{2+}	Mn^{2+}	Fe^{2+}	Co^{2+}	Zn^{2+}	Ni^{2+}	Pt^{2+}	Pd^{2+}	Cu^{2+}	Ag^{2+}
I(volts)	15.03	15.64	16.18	17.05	17.96	18.15	18.54	19.42	20.29	21.48
r(Å)	0.64	0.80	0.74	0.72	0.69	0.69	0.80	0.80	0.70	0.89
U(MO)(kcal/mole)	921	909		957	963	973		969	985	
(MS)	790	796	822 α	850(I)	858sph. 855wz.	863			887	
(MF_2)	696	660	692	714	711	731			727	700
(MCl_2)	597	598	617	644	646	658		661	661	
(MBr_2)	576	582	604	630	633	645		649	645	
(MI_2)	550	563	586	611	617	623			632	

I Increases ⟶

*An even more regular relationship between the crystal energies and I values is noted if compounds regarded as formed from cations of closely similar radius only are considered. In this connection it should be pointed out that in the calculation of crystal energies by the Born-Haber cycle, the term generally of greatest numerical magnitude is the *total* ionization potential leading to the cation.

(4)L. H. Ahrens, *Geochim. et Cosmochim. Acta* 3, 1 (1953).

The fluoride ion, with its small polarizability, appears able to withstand a powerful cationic field, and 6:3 ionic (rutile) structures are found among the fluorides formed from medium-sized divalent cations, even if *I* is large. Rutile structures apparently persist as far as Pd, although some distortion may be present. CuF_2 appears to have a related structure in which, however, the Cu-F distance is anomalously long. The structure of AgF_2 is not known, but is apparently complex. It seems that even F^- is not able to withstand the extremely intense field of the very poorly shielded Ag^{2+} ion. It is noteworthy also, that AgF_2 is the only stable simple salt of Ag^{II}, although stable complexes are known.

In the oxides of Table 14.1, ionic (NaCl) structures persist up to *I* = *c*. 18 V, beyond which structures are non-ionic. The break is not smooth, because whereas zinc is tetrahedrally co-ordinated in ZnO, the structure of NiO has 6:6 coordination; however, the structure of the latter oxide is free from distortion only above 200°C. Square planar coordination is found in PtO, PdO, and CuO (*I* varies from 18.5 to 20.3 V). The oxygen arrangement is most nearly that of a perfect square in PtO, and distortion increases in the sequence PtO-PdO-CuO; that is with increase in *I*.

The sulphide ion S^{2-} is more easily deformed than either F^- or O^{2-}, and a change to non-ionic structures begins at a distinctly lower value of *I* than in the fluorides and oxides. The threshold value seems to be *c*. 15.6 V (*I* for Mn), beyond which only non-ionic structures are found. MnS* itself is trimorphic, one form having 6:6 coordination and two others 4:4 coordination. FeS and CoS have the 6:3 NiAs structure, while ZnS has two dimorphic forms, sphalerite and wurtzite, each with 4:4 coordination. NiS is dimorphic, having a peculiar 5:5 (millerite) structure and also forming the NiAs structure. It seems that Ni^{II} may be located at a threshold value, because if *I* is increased much beyond 18 V only planar structures are found, as in PtS, PdS, and CuS; AgS is unstable. Moreover, although coordination of nickel is not of this type in either dimorph of the sulphide, square planar coordination has been reported in the mineral braggite (Ni, Pt, Pd)S, where Ni^{II} is apparently able to assume this coordination (square planar coordination of S about Ni^{II} may also be found in some complexes, e.g. potassium nickel dithio-oxalate).[5] In the sequence PtS-PdS-CuS, distortion is considerable in the first member, increases in PdS, while in CuS (covellite) a triangle of S atoms replaces the square as nearest neighbours. In the pairs PtO-PtS and PdO-PdS, distortion is greater in the sulphide than in the oxide, whereas in the pair CuO-CuS, only the oxide has planar (distorted) 4:4 coordination. These developments, together with those in the sequence PtO-PdO-CuO, may be summarized as follows:

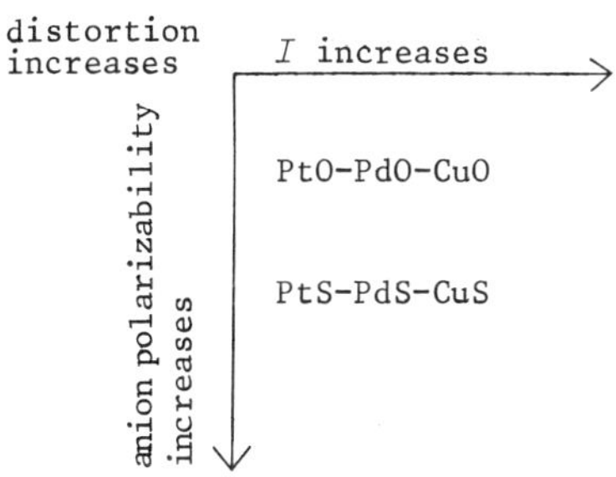

*Polymorphism in MnSe is similar to MnS. MnTe has, however, the NiAs structure, as have FeS, CoS and one of the NiS dimorphs.

(5) E. G. Cox, W. Wardlaw, and K. C. Webster, *J. Chem. Soc.* 1475 (1935).

In chalcopyrite ($CuFeS_2$), coordination of both Cu and Fe is tetrahedral and unlike that in the simple sulphides CuS and FeS. It has been noted[4] that the second ionization potentials of Cu and Fe average at *c.* 18 V, that is, very close to the second ionization potential of Zn; if the average field due to Cu^{2+} and Fe^{2+} has been a significant contribution in determining the coordination of $CuFeS_2$, its structure could be regarded as a compromise structure.

When bonded to radicals ($PO_4{}^{3-}$; $SO_4{}^{2-}$; $VO_4{}^{3-}$, etc.), ionic coordination is found to persist - at least in one compound of this type - all the way from Mg up to and including Cu; H_2O may, however, have to be present in the structure as in $CuSo_4 \cdot H_2O$. Evidently the central atom (P^V, V^V, S^{VI}, etc.) in these radicals binds the oxygen-charge distribution firmly in the radical which is consequently able to withstand the effect of intense cationic fields even with ions such as Cu^{2+}; see also Goldschmidt[6] on contrapolarization. Some of the observations referred to above are summarized in Table 14.2. Arrows in Table 14.2 indicate the approximate range of ionic coordination for given anionic partners.

Compounds regarded as derived from large divalent cations

($r > c.$ 0.95 Å)

The crystal energies of simple crystalline compounds regarded as formed via such cations are shown in Table 14.3. As in Table 14.1, crystal energy varies fairly regularly with I.

Halides. Two significant features about the structures of the halides of the large divalent cations may be noted:

(i) All fluorides, including that of very poorly shielded Hg^{II}, have the 8:4 fluorite structure.* This accords with the radius-ratio rule predictions.

(ii) None of the heavy halides of the large divalent cations has the 8:4 fluorite structure. This comes as a surprise, as some compounds, particularly $BaCl_2$, would be expected to have this structure. A variety of structures is in fact found: the 9-coordinated $PbCl_2$ structure; distorted 6:3 (rutile); $CdCl_2$ and $CdCi_2$ layer-type structures; and ZnS. Evidently the structures of the heavy halides of the large divalent cations are particularly sensitive to such properties as polarizability, cation-anion size relationship, and the effective field of the cation. The apparent significance of these properties may be gauged by examining structural relationships in various groups of halides. In Table 14.4 cation type is kept as uniform as possible by considering alkaline earths only. According to a suggested definition (Ahrens[8]) shielding efficiency is high and about equal in these species. Radius of cation varies considerably, as do the radii and polarizabilities of the anionic halide partners.

The developments in Table 14.4 may be compared with those in Table 14.5. Here one cationic species only is considered, namely Hg^{2+}, in which shielding efficiency is regarded as extremely low (Ahrens[8]), and accordingly the

*PbF_2 is actually dimorphic (fluorite and $PbCl_2$ structures - Wyckoff[7]).

(6) V. M. Goldschmidt, *Geochemistry*, Oxford (1954).

(7) R. W. G. Wyckoff, *Crystal Structures*, Interscience, New York (1951).

(8) L. H. Ahrens, *Nature* 174, 644 (1954).

TABLE 14.2

Crystal structures and approximate ranges of ionic coordination in sulphides, oxides, and fluorides of elements which form medium-sized divalent cations

I increases →

M^{2+}	Mg	Mn	Fe	Co	Zn	Ni	Pt	Pd	Cu	Ag
Sulphides	NaCl	NaCl, ZnS, ZnO (15.6)	NiAs	NiAs	ZnS, ZnO	5:5 millerite, NiAs	coplanar	coplanar	distorted	unstable
	⟶	⟶								
Oxides	NaCl	NaCl	NaCl	NaCl	ZnO (*c*.18)	distorted NaCl	coplanar	coplanar	coplanar	unstable
	⟶	⟶	⟶	⟶						
Fluorides	rutile	rutile	rutile	rutile	rutile	rutile	rutile?	rutile? (*c*.19.5)	fluorite?	?
	⟶	⟶	⟶	⟶	⟶	⟶	⟶ - -	- - →		
Compounds with some polyatomic radicals, some with H_2O									(20.3)	
	⟶	⟶	⟶	⟶	⟶	⟶	⟶	⟶	⟶	

TABLE 14.3

Divalent cations: ionization potentials and crystal energies (U) large-sized (r 0.95–1.35 Å) divalent cations

	Ba^{2+}	Sr^{2+}	Ca^{2-}	Pb^{2+}	Cd^{2+}	Hg^{2+}
I(volts)	10.00	11.026	11.87	15.03	16.904	18.75
r(Å)	1.35	1.18	1.02	1.20	0.97	1.10
U(MO) (kcal/mole)	744	784	831	834 red 834 yellow	903	926 red 926 yellow
(MS)	647	681	724	734	807	848 cinnabar 847 metacin.
(MF_2)	553	589	624	596	662	655
(MCl_2)	483	507	535	534	602	626
(MBr_2)	466	489	515	523	592	620
(MI_2)	442	465	493	511	577	617 red 616 yellow

I increases ⟶

U values have been calculated from experimental thermodynamic data, and refer to 25°C.

TABLE 14.4

Structures of alkaline earth halides (r = radius of cation)

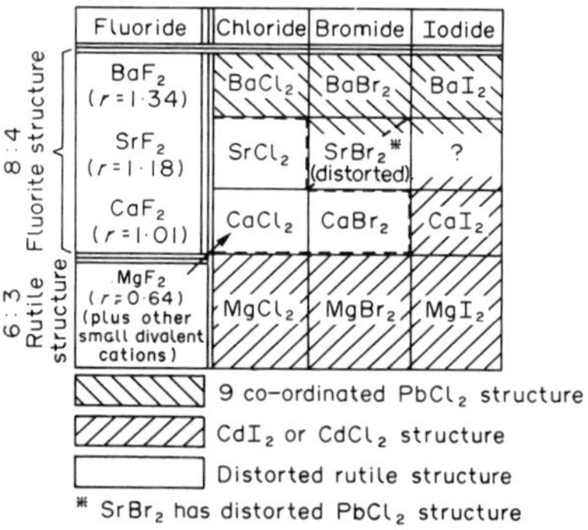

effective field of attraction is extremely high for a cation of this size and charge. The vertical arrangement of compounds is in order of decreasing polarizability of the anion (F < Cl < Br < I). It should be recalled that coordination predicted from ionic-radius ratio rules is 8:4 (fluorite). The fluoride structure is ionic, but those of $HgCl_2$, $HgBr_2$, and HgI_2 (yellow), though ionic in so far that octahedral coordination may be recognized, are severely distorted(2) and in a sense could be regarded as transitional. In this respect the dimorphs HgI_2 (yellow)-HgI^2(red) are of unusual interest, and the relationship may be compared with that of polymorphic MnS (Table 14.2) and of dimorphic HgS (see below). The Hg-I distances of the four tetrahedral neighbours in red HgI^2 are equal to those with the two nearest iodines of the distorted octahedron in yellow HgI_2.

TABLE 14.5

Structure-type and internuclear distances in mercuric halides

Salt	Structure type	Internuclear distances (metal-non-metal)
HgF_2	fluorite	Hg { 8 equidistant *F* atoms at 2.40 Å
$HgCl_2$	distorted octahedral	Hg { 2Cl 2.25 Å; 2Cl 3.34; 2Cl 3.63
$HgBr_2$	distorted octahedral	Hg { 2Br 2.48 Å; 4Br 3.23
HgI_2 (yellow)	distorted octahedral	Hg { 2I 2.78 Å; 4I 4.11
HgI_2 (red)	tetrahedral	Hg { 4 equidistant *I* atoms at 2.78 Å

Wells[(2)] (p. 116) has pointed out that the distorted octahedral structure of $HgBr^2$ may be considered as in between a layer ($CdCl_2$ or CdI_2) and a molecular structure, and this observation is of particular interest when the structural changes in the halides of the divalent metals, as arranged below, are considered.

I increases ↓	Ca^{2+} (r = 1.02 Å)	CaF_2 (fluorite) →	$CaCl_2$, $CaBr_2$ (deformed rutile) →	CaI_2 (CdI_2 layer structure)
	Cd^{2+} (r = 0.97 Å)	CdF_2 (fluorite) →	$CdCl_2$, $CdBr_2$ ($CdCl_2$ layer structure) →	CdI_2 (CdI_2 layer structure)
	Hg^{2+} (r = 1.12 Å)	HgF_2 (fluorite) →	$HgCl_2$, $HgBr_2$ HgI_2 (yellow) (distorted octahedral) →	HgI_2 (red) (tetrahedral)
I increases ↓	Sr^{2+} (r = 1.18 Å)	SrF_2 (fluorite) →	$SrCl_2$, $SrBr_2$ (distorted rutile) →	SrI_2 (structure unknown)
	Pb^{2+} (r = 1.20 Å)	PbF_2 (fluorite and $PbCl_2$ structure) →	$PbCl_2$, $PbBr_2$ ($PbCI_2$ structure) →	PbI_2 (CdI_2 layer structure)

It may seem a little surprising that most of the heavy halides of mercury are structurally ionic, albeit coordination is distorted, and this aspect is considered in a little detail below in the discussion on the oxides and sulphides of the divalent metals with large cationic radii.

Structures of sulphides of metals which form large divalent cations

Sulphides are listed below, the horizontal arrangement being in order of increasing second ionization potential of the electropositive constituents. Ionic radii of the corresponding cations are also given.

I increases ⟶

	BaS	SrS	CaS	SnS	PbS	CdS	HgS
r(Å)	1.35	1.18	1.02	0.93	1.20	0.97	1.10

BaS, SrS, and CaS have ionic (NaCl) structures. SnS has a distorted NaCl structure and PbS has the NaCl arrangement, whereas coordination in CdS is non-ionic (tetrahedral). As already noted, HgS is dimorphic, coordination in the black variety (metacinnabarite) is tetrahedral whereas that in the red (cinnabar) is octahedral.

The presence of ionic coordination in cinnabar, albeit distorted, and in PbS (galena), may be due to the large size of the Hg^{2+} and Pb^{2+} ions. It seems otherwise surprising that as coordination in CdS is tetrahedral, that in cinnabar could turn out to be octahedral, because as far as it can be ascertained, the effective field associated with Hg^{2+} is considerably greater than that of Cd^{2+}. The inference is that, had the Hg^{2+} radius been slightly smaller, say 1.05 Å as compared with the estimated 1.10 Å, only tetrahedral coordination would obtain in HgS, as is found in metacinnabarite. The presence of ionic coordination in galena can perhaps be interpreted in the same way. It may be assumed, as before, that reaction to form PbS proceeds via ions (Pb^{2+} and S^{2-}) and that because the effective force of attraction associated with Pb^{2+} is quite high - as compared with Sr^{2+} of similar size, for example - the charge distribution about S^{2-} is seriously deformed or polarized. The resultant PbS is consequently described as predominantly covalent rather than ionic, but, because Pb^{2+} is very large, a complete rearrangement of electron distribution, involving a collapse to lower non-ionic coordination, is prevented from taking place.

Although coordination in PbS is not seriously deformed, that in $PbSnS_2$ (teallite) is distorted and similar to that in SnS. This relationship differs from that in the non-ionic sulphides CuS, FeS, $CuFeS_2$, and ZnS referred to above. Pb and Sn have fairly similar *I* values compared with Cu and Fe, where the values are quite widely different. If the average field of the two metal ions is important for determining the coordination in a compound with two electropositive constituents (see p.), it can be readily understood why the coordination in the compromise structure of $PbSnS_2$ should resemble that of SnS and PbS, whereas the compromise structure of $CuFeS_2$ bears little resemblance to the structures of CuS and FeS.

Crystalline compounds of univalent metals

The structures of simple compounds of univalent metals are shown in Table 14.6 in order of first ionization potential of the metal constituents.

Changes of structure with increase in *I* cannot be examined quite so critically in compounds of the univalent elements, because univalent cations are relatively easily polarized and this property may also influence structure (see below). Nevertheless, in general terms, we may observe that structure usually changes from ionic to non-ionic with increase in *I* (and with increase in the polarizability of the anion). Gold is particularly interesting in this respect. As defined,[8] the shielding efficiency of the Au core in Au^+ is lower than in any other univalent cationic core and hence the effective field strength associated with Au^+ is very great indeed. This may be a reason why no simple Au^I compound is structurally ionic; the most likely possibility, AuF, is apparently not well characterized. (Extremely inefficient shielding in gold seems to account plausibly for some of the unusual chemistry of this element - extreme inertness, low atomic volume, extreme tendency to form complexes in solution, complete absence of structurally ionic compounds, and so on).

TABLE 14.6

Structures of simple crystalline compounds of univalent metals

	Cation	*r*	Fluorides MF	Chlorides MCl	Bromides MBr	Iodides MI	Oxides M_2O	Sulphides M_2S
	Cs^+	1.65 Å	NaCl	CsCl	CsCl	CsCl	$CdCl_2$	complicated
I increases ↓	Rb^+	1.44	NaCl	NaCl	NaCl	NaCl	CaF_2	CaF_2
	K^+	1.33	NaCl	NaCl	NaCl	NaCl	CaF_2	CaF_2
	Na^+	0.95	NaCl	NaCl	NaCl	NaCl	CaF_2	CaF_2
	Tl^+	1.47	distorted NaCl	CsCl	CsCl	CsCl		complicated
	Ag^+	1.26	NaCl	NaCl	NaCl	ZnS, ZnO	Cu_2O	acanthite (low)
	Cu^+	0.91	ZnS	ZnS	ZnS	ZnS	Cu_2O	orthorhombic (low)
	Au^+	1.37						

For details of these structures, see Wyckoff[7]

RO_2 STRUCTURES

Coordination in the oxides of the quadrivalent elements invariably conforms quite closely to that predicted by the radius-ratio rules. There is some evidence, however, that the effective field of the cation might influence structure. Consider the following sequence of M^{4+} cations, arranged in order of increasing radius -

	... Mo^{4+},	Sn^{4+},	Nb^{4+},	Hf^{4+},	Zr^{4+},	Pb^{4+},	Am^{4+},	Ce^{4+},	U^{4+} ...
radius (Å)	(0.70)	(0.71)	(0.74)	(0.79)	(0.79)	(0.85)	(0.92)	(0.94)	(0.97).

The oxides of cations with radii *c*. 0.70 Å have 6:3 (rutile) coordination, whereas oxides of the large cations have the 8:4 fluorite structure. The trio Hf-Zr-Pb are in a sense located in a transition zone. In the pair HfO_2-ZrO_2, the first component has the fluorite structure, whereas ZrO_2 has a distorted structure at room temperature and forms the fluorite structure only at elevated temperatures. Although Zr and Hf have very similar properties, the fourth ionization potential of Zr (*c*. 34 V) is slightly but distinctly greater than that of Hf (*c*. 31 V). Also, although the radius of Pb^{4+} is apparently a little greater than that of Hf^{4+}, PbO_2 has the rutile structure, and this could perhaps be caused in part by the fact that the fourth ionization potential of Pb (39.0 V) is distinctly elevated.

THE FLUORIDES OF K, Rb, Tl, AND NH_4

Discussion so far has dealt primarily with the possible significance of the effective field of a cation for determining structure, and the polarizability of the anion has been regarded as an important property. In this section some evidence of structural change which seems to be correlateable with deformability of the *cation* will be discussed. We recall the general rule that cations are most easily polarized in the field of the anion when they are large and univalent. Maximum cation deformation is apparently caused by F^- (Fajans; several papers). Consider now the fluorides of K^+, Rb^+, Tl^+, and NH_4^+ (r varies from 1.33 to 1.44 Å); cation polarizability probably increases in the order, K^+, Rb^+, Tl^+, and NH_4^+. The structures of KF and RbF are ionic (NaCl), with little apparent distortion; coordination in TlF is still octahedral, but there is severe distortion, and instead of six equidistant F neighbours about Tl, the distribution is 2 (2.59 Å), 2 (2.75 Å), and 2 (3.04 Å). Such distribution is very similar to that in the one HgS dimorph - see above. NH_4F has the wurtzite structure with 4:4 coordination.* Structural changes in the sequence RbF-TlF-NH_4F may be compared with those in the halides of mercury, thus:

Octahedral	Distorted octahedral (2:2:2)	Tetrahedral
RbF	TlF	NH_4F
HgF_2	$HgCl_2$	HgI_2
	(HgS cinnabar)	(HgS metacinnabar)

In each sequence, similar developments take place, but in the one, variable deformation of the cation may have been the cause, and in the other, variable polarization of the anion.

Some general observations

The data considered above provide quite distinct evidence of the existence of ionization potential threshold values which may serve to divide ionic from non-ionic structures. Within the non-ionic structure range there is evidence also of variation of structure with ionization potential. A question arising from this observation is whether extreme anion polarization deformation, involving perhaps serious rearrangement of electron distribution, might itself in part determine or at least influence bond disposition in non-ionic inorganic compounds. It is one thing to draw attention to the existence of certain regularities and discontinuities, but another to develop a satisfactory theory to account for them. Such an attempt will not be made here, but attention will be drawn to some structures which have been explained on the basis of extreme anion polarization and not resonance hybridization. Such a hypothesis might ultimately serve as a useful model for basing a more general theory.

The structures of NaOH, KOH, RbOH, and CsOH are ionic (NaCl), whereas the dihydroxides of the alkaline earths and Cd, Mn, Co, Ni, and Zn (one dimorph only) are not ionic (fluorite or rutile, according to the size of the cation) but have the CdI_2 structure (compare with chlorides and bromides of the

*This is usually attributed to hydrogen bonding.

divalent metals). The hydroxides of the trivalent metals (Fe, Al, Cr?) with fairly small cations, contain hydroxyl bonds. Bernal and Megaw[9] have suggested that the ion OH^- undergoes varying degrees of polarization, depending upon the field of the cation; polarization may be extreme, producing a separation of charges arranged tetrahedrally, and ultimately migration of H from one O to another. Change of structure and departure from simple ionic structures is suggested to arise from varying OH^- ion polarization and hence not from some form of resonance hybridization. It might be possible to adopt such a scheme for various non-ionic structures. Each anion will respond to the field of the cation in a different way; we have sufficient evidence to indicate that some anions are more polarizable than others, but we are not informed about the ways in which different anions will respond as the field of the cation becomes greater.

---

[9] J. D. Bernal and H. D. Megaw, *Proc. Roy. Soc.* A151, 384 (1935).

CHAPTER 15

Variation of Refractive Index with Ionization Potential in Some Isostructural Crystals

Many authors have discussed variation of refractive index with composition: Barth (1930), Hess (1949), Kennedy (1947), Hori (1954 and 1956), Jaffe (1956), and Larsen (1921 and 1934) are a few examples in mineralogy. The present paper is also concerned with the general problem of variation of refractive index with composition: its main purpose will be to seek a closer understanding of the physical causes of the variation of refractive index in ionic crystals.

The refractive index of a crystal depends mainly on its structure (Hori, 1954 and 1956, for example) and certain intrinsic properties of its component atoms or ions (Fajans and Kreidl, 1948, who refer to several earlier publications of Fajans and co-workers). If we consider isostructural ionic crystals only and confine ourselves to examining *variation* of refractive indices in suites of such crystals (the olivines, for example), it is necessary to consider only the properties of the ions in question. Under these conditions and for the purpose of this study the refractive index itself rather than the molar refraction will be used.

In ionic isostructural crystals, variation of refractive index should reflect the average state of polarization of the constituent cations and anions; the greater the degree of polarization (deformation), the greater the interaction with the passage of light waves and the higher the refractive index. The familiar generalization that anions are much more easily polarized (deformed) than cations may be assumed to be valid provided large and singly charged cations (Na^+, K^+, Rb^+, Cs^+, NH_4^+, Tl^+, and Ag^+) are omitted. Accordingly, in crystals other than those involving large singly charged cations, polarization of the anion should make the major contribution to the total refractive index because polarization of the anion will be far greater than that of the cation. The degree of polarization of the anion depends on the field in which it finds itself and this is controlled largely by the effective positive force of attraction of the cation. This force of attraction (polarizing power) depends on the charge and size of the cation and the efficiency with which its electrons shield (screen) the nuclear charge. A further contributing property may be the response of certain $3d$ transition metal cations to the field of the crystal — so-called crystal-field splitting.

The ionization potential may be used to indicate the effective force of attraction of the cation (see, for example, Ahrens (1953) and Ahrens and Morris (1956)); the greater the energy needed to remove an electron the greater the polarizing power of the resultant cation. In suites of isostructural ionic crystals, those that contain cations of metals with the highest ionization potentials should have the highest refractive indices because the constituent anions should be deformed to the greatest extent. This prediction may be tested by examining groups of isostructural crystals, but because of a number of variables the comparisons should be made as rigorous as possible and in each of the discussions which follow, the anion will be the same, cationic charge will be constant, and the radii of the cations will be similar. Cationic charge distribution is another factor: whereas the charge distribution on 8- and 18-electron cations is probably spherical, that on the transition cations may be different and accordingly transition cations are omitted from the first comparisons.

For our purpose, an ionic compound is taken to mean one in which the ionic radius ratio rules are obeyed: this is not meant to imply that the bonds are regarded as purely ionic.

Refractive indices are usually either from Winchell (1931) or Winchell (1951).

ISOSTRUCTURAL IONIC COMPOUNDS INVOLVING 8- AND 18-ELECTRON CATIONS: Al^{3+} AND Ga^{3+}; Si^{4+} and Ge^{4+}; Ca^{2+} and Cd^{2+}; Mg^{2+} AND Zn^{2+}

The α refractive indices of the Ga and Ge substituted felspars (Goldsmith, 1950) are: $KAlSi_3O_8$, α 1.518; $KGaSi_3O_8$, α 1.533; $KAlGe_3O_8$, α 1.595; $KGaGe_3O_8$, α 1.615. Elevation of refractive indices by replacement of Al^{3+} and Si^{4+} by Ga^{3+} and Ge^{4+} respectively accords with prediction. The third ionization potentials of Al and Ga are 28.5 and 30.7 volts respectively, and the fourth ionization potentials of Si and Ge are 45.1 and 45.7 volts respectively. In each pair, the force of attraction of the 18-electron cation is evidently greater than that of the 8-electron cation and should cause greater polarization of the anion. The example is quite a good one as radii are similar; Si^{4+} = 0.42 Å, Ge^{4+} = 0.47 Å: Al^{3+} = 0.51 Å, and Ga^{3+} = 0.57 Å.

CaF_2 and CdF_2. The radii of the two cations, Ca^{2+} and Cd^{2+}, are very similar, namely 1.01 and 0.97 Å. Their second ionization potentials (11.9 and 16.9 volts respectively) differ quite considerably (5 volts) and large refractive index differences are anticipated. Unfortunately structurally ionic compounds of cadmium are rare and few comparisons are possible. The pair, CaF_2-CdF_2, each having the 8:4 fluorite structure, is one good example. Goldschmidt (1926) observed that CaF_2 and CdF_2 were completely miscible and that refractive index varied regularly with composition; for CaF_2, n = 1.43 and for CdF_2, n = 1.56. The difference is quite large and accords with theory. The refractive index difference would probably be much larger if the anion were easily polarized.*

In the isostructural pair CaO and CdO (both NaCl-type), the respective refractive indices (average for various wavelengths) are 1.84 and 2.49; the difference is very large.

---

*F^- is the least polarizable of all anions and this appears to account for the fact that the refractive indices of fluorides are often conspicuously low; compare fluor- and hydroxy-apatites, for example.

Mg and Zn compounds. The structures of many zinc compounds are not ionic but where comparisons can be made, the refractive indices of the zinc compounds are greater than those of magnesium. This accords with theory; for Zn, I_2 = 17.9 volts and for Mg, I_2 = 15.0 volts (Zn^{2+} = 0.69 Å and Mg^{2+} = 0.65 Å). Further discussion on the Mg-Zn relationship is deferred to a later section which compares variation of refractive indices in isostructural compounds involving a variety of medium-sized cations.

ISOSTRUCTURAL IONIC COMPOUNDS INVOLVING SOME TRANSITION ELEMENT CATIONS

Refractive index comparisons between compounds of the pairs of Group IV, V, and VI elements, Zr—Hf, Nb—Ta, and Mo—W, respectively, are particularly satisfactory because for each pair the respective cationic radii are virtually the same and electron configurations are very similar.

In the pair Zr and Hf (radius of Zr^{4+} = Hf^{4+} = 0.79 Å), the fourth ionization potentials are ~34 and ~31 volts, respectively. The refractive indices of Zr and Hf compounds are therefore predicted to be very similar, with those of Zr slightly greater than those of Hf. In the isometric compounds $(NH_4)_3ZrF_7$ and $(NH_4)_3HfF_7$, the respective refractive indices are 1.433 and 1.426, which accords with theory. Zirconium and hafnium form tetragonal oxyhalides; $ZrOCl_2.8H_2O$ is uniaxial positive (ω = 1.552, ε = 1.563) whereas $HfOCl_2.8H_2O$ is uniaxial negative (ω = 1.557, ε = 1.543). In this example the comparison is not so clear: the refractive indices are closely similar as we expect and if either the average or $\sqrt[3]{\omega^2\varepsilon}$ is compared, the value for the Zr compound is slightly greater, which is in accordance with theory.

For the pair Nb^{5+} and Ta^{5+} (radius = 0.69 Å) the fifth ionization potentials are ~50 and ~45 volts respectively. A difference of about 5 volts for quintuply charged cations is comparatively small, and accordingly we predict that the refractive indices of isostructural compounds of niobium and tantalum will be similar but that those of niobium will be slightly greater. The refractive indices of stibiocolumbite and stibiotantalite support this prediction:

$SbNbO_4$	α 2.42	β 2.44	γ 2.46
$SbTaO_4$	α 2.345	β 2.383	γ 2.452.

In tantalite and columbite $(Fe,Mn)(Nb,Ta)_2O_6$ Fe—Mn substitution is a complication. The refractive indices of columbite are, however, greater than those of tantalite.

In the pair Mo^{6+}—W^{6+} (radius = 0.62 Å), the sixth ionization potentials are ~67 volts and ~60 volts. Refractive indices of the isostructural tetragonal pairs powellite-scheelite and wulfenite-stolzite are as follows:

$CaMoO_4$	ω 1.967	ε 1.978	$PbMoO_4$	ω 2.405	ε 2.283
$CaWO_4$	ω 1.918	ε 1.934	$PbWO_4$	ω 2.27	ε 2.19.

The refractive indices of the Mo member in each pair are greater than those of the W member, which is in accordance with theory. We may note also that the refractive indices of the lead minerals are quite distinctly greater than their calcium equivalent (Pb^{2+} = 1.20 Å, Ca^{2+} = 1.01 Å). This also accords with theory as the second ionization potentials of lead and calcium are 15.0 and 11.9 volts respectively; the comparison is, however, not rigorous, as the difference in radius is quite considerable (0.2 Å) and a large divalent cation such as Pb^{2+} is itself perhaps quite strongly polarized in the anion field, thereby causing a further elevation of refractive index.

MEDIUM-SIZED DIVALENT CATIONS OF VARIOUS TYPES

The largest number of cations of a similar size is the medium-sized group Mg^{2+}, Mn^{2+}, Fe^{2+}, Co^{2+}, Ni^{2+}, Cu^{2+}, and Zn^{2+}; their radii range from 0.65 Å (Mg^{2+}) to 0.80 Å (Mn^{2+} and perhaps Cu^{2+}). It would be desirable to include Pd^{2+} and Pt^{2+} also, but refractive index data on their compounds are not often available. Though the exclusion of other divalent cations from the above series is somewhat arbitrary, we may recall that the fluorides of the cations of this group invariably have the 6:3 rutile structure, whereas that of BeF_2 (Be^{2+} = 0.34 Å) is 4:2 (cristobalite) and the fluoride structures of many larger cations (Ca^{2+}, Sr^{2+}, Ba^{2+}, Ra^{2+}, Cd^{2+}, and Hg^{2+}) are 8:4 (fluorite).

The second ionization potentials in volts of the metals under consideration are Mg 15.0, Mn 15.6, Fe 16.5, Co 17.4, Ni 18.2, Cu 20.2, and Zn 18.0. In two types of compounds it is possible to make a comparison involving each of the seven cations. Consider the trigonal hydrated divalent-quadrivalent hexafluorides, $A^{2+}SiF_6.6H_2O$, where A^{2+} may be Mg, Mn, Fe, Co, Ni, Cu, or Zn. The Mg, Mn, Fe, Co, Ni, and Zn compounds are uniaxial positive whereas the Cu compound is uniaxial negative.

The second ionization potential is related to refractive index in Fig. 15.1. A well-developed regularity is clear and in general the relationship is in accordance with theory outlined so far. The change of sign in the copper compound might reflect some unique property associated with Cu^{2+} and in the example below (see Fig. 15.2) the copper compound again shows a peculiarity. It may be recalled that although coordination about Cu^{2+} can often be regarded as octahedral, four neighbours are usually closer than the other two. The Mg—Fe relationship above holds generally for mafic minerals such as olivine and pyroxene and the higher refractive indices of the Fe-rich component are probably due to greater deformation of O^{2-} in the field of Fe^{2+}.

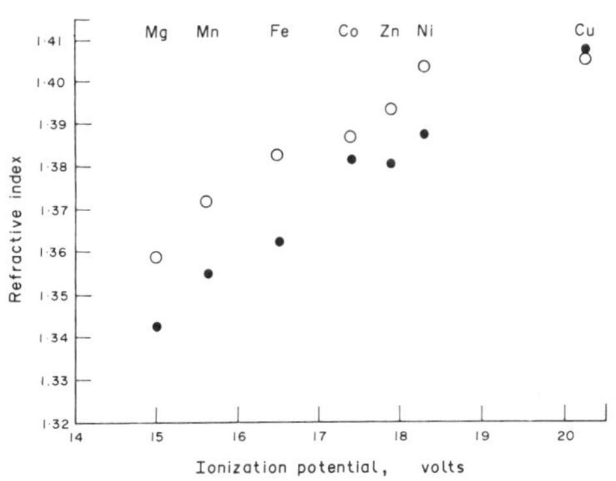

Fig. 15.1. Variation of refractive indices with second ionization potential of metal in tetragonal crystals of the type $A^{2+}SiF_6.6H_2O$. • = ω, O = ε.

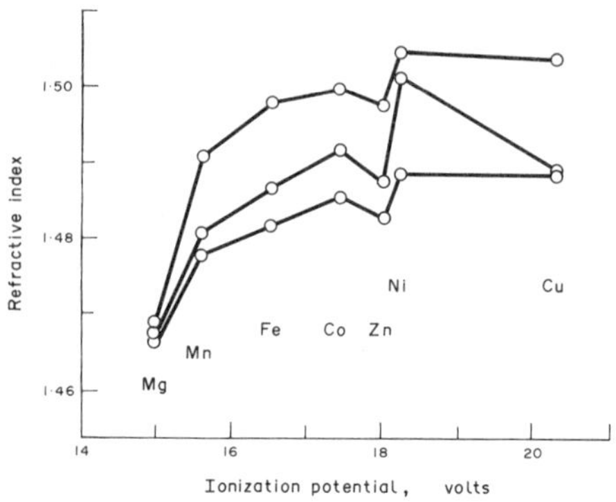

Fig. 15.2. Variation of refractive indices with second ionization potential of metal in monoclinic sulphates of the type $Rb_2B^{2+}(SO_4)_2.6H_2O$.

A large number of compounds of the type $A_2^+B^{2+}(SO_4)_2.6H_2O$ are known in which A^+ is a large singly charged cation (K^+, Rb^+, Cs^+, Tl^+, or NH_4^+) and B^{2+} is a medium-sized divalent cation. The relationship between second ionization potential and refractive index shown in Fig. 15.2 refers to the rubidium compounds but holds fairly well for the others as well. The general trend of Fig. 15.1 is again clear in Fig. 15.2, but the regularities are not so well developed and distinct discontinuities appear at zinc. Also, unlike the fluosilicates, the refractive indices of copper rubidium sulphate tend to be less than those of nickel, particularly the β index. In the K, Cs, NH_4, Tl salts, as well as the isomorphous selenates (the Cs selenate excepted), the drop in β index is smaller when passing from the nickel compound to the copper compound.

We may note one difference between the Tutton salts and the fluosilicates that may be important: in the Tutton salts each cation is surrounded by six H_2Os which form an octahedron distorted from the regular shape by being elongated along an axis passing through two opposite water molecules, but having very nearly tetragonal symmetry (Hoffman, 1931; Abragam and Pryce, 1951); magnetic experiments indicate a crystal electric field of tetragonal symmetry. In cobalt fluosilicate (Abragam and Pryce, 1951) and perhaps also in the other fluosilicates, the octahedral group of H_2O molecules is more nearly regular than in the Tutton salts; distortion is in the form of compression along a trigonal axis and the cation is located in a crystal electric field of trigonal symmetry. A refractive index drop at zinc, as above, may be observed also in some other compounds such as the carbonates and simple sulphates, and in fact the relationship $Mg^{2+} < Zn^{2+} < Fe^{2+}$, Co^{2+}, Ni^{2+}, and Cu^{2+} is quite common; thus the refractive index of the compound with an 8-electron cation is lowest, that containing the 18-electron cation comes next, whereas compounds containing 3*d* transition element cations may have the highest refractive indices. Such a relationship holds quite often also for compounds of medium-sized R^{3+} cations: Al^{3+} (8-electron) < Ga^{3+} (18-electron) < Cr^{3+}, Fe^{3+}, Co^{3+}, and Rh^{3+} (all transition). In the alums, for example, the relationship is Fe^{3+} (highest) > Ga^{3+} > Al^{3+}, and the same relationship holds also for orthoclase where synthetic Fe^{3+} orthoclase has α and γ 1.601 and 1.609, respectively (Gaubert, 1926; see also Coombs, 1954), compared with lower values for Ga and still lower for Al orthoclase (see above). These relationships accord only in part with the magnitudes of the ionization potentials and lack of complete correlation, when a variety of cation types are considered, may be due to the effect of the crystal-field on the transition element cations.

REFERENCES

Abragam, G. and Pryce, M. H. L. (1951) *Proc. Roy. Soc.*, ser. A, 206, 173.
Ahrens, L. H. (1953) *Geochimica Acta* 3, 1.
Ahrens, L. H. and Morris, D. F. C. (1956) *Journ. Inorg. Nucl. Chem.* 3, 270.
Barth, T. F. W. (1930) *Amer. Journ. Sci.*, ser. 5, 19, 143.
Coombs, D. S. (1954) *Min. Mag.* 30, 409.
Fajans, K. and Kreidl, N. J. (1948) *Journ. Amer. Ceram. Soc.* 31, 105.
Gaubert, P. (1926) *Compt. Rend. Congrès Soc. Sav.*, 402.
Goldschmidt, V. M. (1926) *Skrifter Norske Vidensk.-Akad., Mat.-nat. Kl*, No.2.
Goldsmith, J. (1950) *Journ. Geol.*, Chicago, 58, 518.
Hess, H. H. (1949) *Amer. Min.* 34, 621.
Hofmann, W. (1931) *Zeits. Krist.* 78, 279.
Hori, F. (1954) *Sci. Papers Coll. Gen. Educ., Tokyo*, 4, 71.
Hori, F. (1956) *Min. Journ.* [Japan], 1, 359.
Jaffe, H. W. (1956) *Amer. Min.* 41, 757.
Kennedy, G. C. (1947) *Ibid.* 32, 561.
Larsen, E. S. (1921) *U.S. Geol. Surv.*, Bull. 679, 30.
Larsen, E. S. and Berman, H. (1934) *Ibid.*, Bull. 848, 30.
Winchell, A. N. (1931) *The Microscopic Characters of Artificial Inorganic Solid Substances or Artificial Minerals.* New York (John Wiley).
— and Winchell, H. (1951) *Elements of Optical Mineralogy*, Part 2, New York (John Wiley).

CHAPTER 16

Geochemistry

This chapter is divided into four sections:

(i) Siderophile tendency.
(ii) Separation of elements into sulphides (chalcophile tendency) and silicates (lithophile tendency).
(iii) The apparent affect of *IP*s and covalency on element distribution in igneous rocks and minerals, and
(iv) Notes on the enrichment of some metals in the organic fraction of sediments.

SECTION (i)

Elements which tend to concentrate in metallic iron - whether in the metal phase of meteorites; in the mineral, native iron; in a possible Fe-rich core of the earth; or perhaps in pig iron as obtained from a blast furnace - are commonly referred to as siderophile (Goldschmidt). Defined thus, a siderophile tendency is not the same as the tendency of a metal to occur in the native state in any geological environment; bismuth may, for example, occasionally occur in the native state and if so is virtually free from iron. Nevertheless, siderophile tendency and a tendency to occur in the native state are closely related, in particular in magmatic and meteoritic environments, where strongly siderophile elements are typically those which have a decided tendency to occur in the native state; Au and the platinum group elements are examples. Such elements tend to be chemically inert and for the sake of gaining greater insight into the causes for a siderophile tendency and the related general tendency to occur in the native state it is quite fruitful to consider the causes of chemical inertness (Chapter 9).

If firmness of binding, as indicated by the ionization potential, determines to some considerable extent the tendency of an element to be inert, some correlation between *IP* and siderophile tendency may be expected if elements are arranged in order of their first ionization potentials. Such an arrangement is shown below and refers to all elements which may be univalent in geological environments.

Four ionization potential magnitudes may be recognized, as indicated by the "boxing". In general, the tendency to be less reactive increases from left to right with Au (I_1 = 9.2 eV) the most inert of the univalent elements and showing a particularly strong tendency to be siderophile and occur in the uncombined state. In fact, Vincent and Crocket (1960) have suggested that gold may occur in silicate minerals and rocks as dispersed uncombined neutral atoms. Ag and Cu can be siderophile but to a lesser degree than Au;

TABLE 16.1

Cs, Rb, K, Na, Li	Tl	Ag, Cu	Au
(3.9 - 5.3)	(6.1)	(7.5 - 7.7)	(9.2)

IP increases →

The extreme firmness of binding of the outermost electron in gold is anomalous (Chapter 9) because the ionization potential increases when passing from Ag to Au, instead of decreasing. This elevation in *IP* is evidently due to the effect of entry of the $4f$ and/or $5d$ electron (Ahrens, 1953, and Chapter 9) and were it not for this anomaly, Au would have been considerably more reactive, and would probably have existed in a combined state in geological environments: its chemistry and geochemistry would probably have been midway between that of Ag and Tl. (It is of interest to note that were it not for the anomaly, the world's economy and our life styles would have been very different.)

When considering the Group II elements or other elements which are usually divalent, it is not clear as to whether the first or second ionization potentials or perhaps both should be used as a guide to chemical inertness and siderophile tendency. Ahrens (1953) used the second ionization potentials; these are given below for the Group II elements.

Be ($2s^2$)	18.21 eV		
Mg ($3s^2$)	15.03 "		
Ca ($4s^2$)	11.9 "	Zn ($4s^2$)	17.96 eV
Sr ($5s^2$)	11.03 "	Cd ($5s^2$)	16.9 "
Ba ($6s^2$)	10.00 "	Hg ($6s^2$)	18.75 "

The IIb elements Zn, Cd and Hg are generally more chemically inert than the IIa elements; compare particularly the pairs Ca-Zn, Sr-Cd and Ba-Hg in each of which the principal quantum number is the same. The relatively higher ionization potential in the Group IIb elements indicates a much greater firmness of binding of the s electrons and hence a greater tendency to be chemically inert and exist in the native state. Neither element is, however, siderophile. As in the Group 1b elements where Au is particularly inert, Hg (the element following Au) is the most inert of the Group 2b elements and has by far the highest second ionization potential. Native Hg exists and Hg amalgam minerals are known (Pd, Hg (potarite) for example) but Hg is, however, not siderophile as it is not associated with Fe, but tends to be chalcophile.

If the $3d$ transition metals alone are considered, inertness and siderophile tendency follow ionization potential trends quite closely. Thus, in the sequence Mn (d^5s^2), Fe (d^6s^2), Co (d^7s^2) and Ni (d^8s^2) siderophile tendency increases (Mn to Ni) with firmness of binding as indicated by the second ionization potentials (15.64, 16.18, 17.05 and 18.15 eV respectively). Strongly siderophile Pt ($5d^9 6s^1$) and Pd ($5d^{10}$) have second ionization potentials (16.6 and 19.42 eV respectively), greater than moderately siderophile Ni. If, however, all types of divalent elements are considered, correlations are not quite as good. Nevertheless, it is clear that ionization potentials provided a meaningful measured guide - through the idea of firmness of binding and chemical inertness - as to why many metals tend to occur in the native state and have siderophile tendencies.

Concerning lithophile, chalcophile and siderophile tendencies, Mason (1958) states: "Brown and Patterson have shown that if the heat of formation of an oxide is greater than that of FeO the element is lithophile; the difference between the two heats of formation is a measure of the intensity of the lithophile character. Similarly, those elements having oxides with heats of formation lower than FeO are chalcophile or siderophile; those for which the heat of formation of the sulfide is equal to or greater than that of FeS are generally chalcophile. A semi-quantitative measure of lithophile, siderophile, or chalcophile character is also provided by the electrode potential. Elements with high positive potentials (1 - 3 V), such as the alkali and alkaline earth metals, are lithophile; the noble metals, with high negative potentials, are siderophile; elements falling in the intermediate range are generally chalcophile." It should be noted that the heat of formation is related to the ionization potential.

SECTION (ii)

The metal-sulphur bond in most sulphide minerals is usually classed as mainly covalent whereas the bonds between oxygen and the abundant metals (Na, K, Ca, Mg, Fe, Al and Si) in silicate and other oxyminerals are usually considered as mainly ionic because coordination in these minerals accords with the requirements of the ionic radius ratio rules.

Sulphides and silicates are not miscible and such lack of miscibility is evidently due to a significant difference in bond character (Ahrens, 1953). These differences evidently arise in the first place from differences in the properties of the atoms of the two non-metals in question, oxygen and sulphur, and secondly from differences in the properties of the metals.

For the purpose of further discussion it will be assumed that regardless of whether the bond in a mineral is ionic or not, many reactions proceed via ions (e.g. $Zn^{2+} + S^{2-} \rightarrow ZnS$ (covalent solid)) and the properties of *ions* are therefore stressed. Consider the sulphides and oxides of the elements Mg and Fe. The respective Mg—S and Fe—S bonds will be distinctly more covalent than the corresponding Mg—O and Fe—O bonds. Whether, however, an element will actually have a chalcophile tendency appears to depend on whether the degree of covalency in the bond is sufficiently high. In the above pair Fe has both a lithophile and a chalcophile tendency whereas Mg shows a marked lithophile tendency only. Clearly the atoms of Mg and Fe differ in one or more important respects. The effective force of attraction (polarizing power) of the cation is probably the most important single property (Förland and Weyl, 1949; Ahrens, 1953; Goldschmidt, 1954). The sulphides of some metals are structurally ionic (for example, MgS and the other alkaline earth sulphides and one dimorph of MnS), but in the mineral sulphides the structures of which are not ionic, the metals are invariably those whose cations (given charge and size) have high polarizing powers. Accordingly, we will arrange cations (given charge and radius) in order of increasing polarizing power, as measured by the *n*th ionization potential in order to ascertain whether threshold values of polarizing power may exist which, if exceeded, give rise to sulphide bonds of such covalency that the metals can be accepted into sulphide minerals (Ahrens, 1953).

Univalent metals

Cations of given radius are arranged (Table 16.2) in order of increasing first ionization potential.

TABLE 16.2

Univalent cations

	Rb^+	K^+	(Na^+)	Tl^+	Ag^+	(Cu^+)	Au^+
I_1 (eV)	4.2	4.34	5.14	6.1	7.57	7.7	9.22
Radius (Å)	1.45	1.33	0.97	1.45	1.27	0.97	1.37

The ions Na^+ and Cu^+ are bracketed because their radii are distinctly smaller than the radii of the other cations and are therefore not strictly comparable. Li^+ (0.68 Å) and Cs^+ (1.67 Å) have been omitted for the same reason.

It is evident that if a univalent metal has a first ionization potential greater than about 6 eV, the bond it forms with sulphur will have a sufficiently high degree of covalent character to enable it to either form a sulphide mineral or be accepted into the covalent sulphide mineral structures. Thallium sits, in a sense, neatly on the fence as its chemical properties are similar to K^+ and Rb^+, and Ag^+. A most interesting geochemical feature of thallium arises from the fact that Tl is associated with K and Rb in the ionic silicates (see iii) yet this element, but not the alkali metals, is also accepted into the structures of various sulphide minerals.

Electronegativity data support in a general way the principal conclusion (the presence of a threshold value) from the ionization potential data. In general, however, the electronegativity data present a much less satisfactory picture (Ahrens, 1964) due evidently to various differences in the calculation of this quantity. For example, one set of data indicate that the electronegativity of Ag is less than Tl and the other the reverse. Further, according to one set of values, Ag and Cu have equal electronegativities whereas the other set indicates that the electronegativity of Cu is distinctly greater than that of Ag. In one set of values the electronegativity of Au is given as less than Cu which seems surprising. For details, see Ahrens (1964).

Divalent cations

Divalent cations are separated into two size groups (Table 16.3). The division is to some extent arbitrary, but is based in part on the fact that the 6:3 rutile structure is common to fluorides of medium-sized group cations whereas the 8:4 fluorite structure prevails in nearly all fluorides of the large-sized cations.

TABLE 16.3

Medium-sized divalent cations (0.65 - 0.8 Å)

	Mg^{2+}	Mn^{2+}	Fe^{2+}	Co^{2+}	Zn^{2+}	Ni^{2+}	Pt^{2+}	Pd^{2+}
I_2 (eV)	15.03	15.64	16.24	17.4	17.94	18.2	19.3	19.9
Radius (Å)	0.65	0.80	0.74	0.72	0.69	0.69	0.80	0.80

Large divalent cations (0.93 - 1.35 Å)

	Ba^{2+}	Sr^{2+}	Ca^{2+}	Sn^{2+}	V^{2+}	Pb^{2+}	Cd^{2+}	Hg^{2+}
I_2 (eV)	10.00	11.03	11.9	14.63	14.65	15.05	16.9	18.9
Radius (Å)	1.34	1.12	1.01	0.93	~0.9	1.20	0.97	1.10

For the medium-sized cations (0.65 - 0.80 Å) the magnitude of the second ionization potential should be greater than about 15.5 eV before the metal-sulphur bond is sufficiently covalent to enter the structures of sulphide minerals. As for thallous thallium, divalent manganese is located on a threshold. Though a fairly abundant element, manganese as Mn^{II} does not form sulphide minerals of its own; perhaps the reasons are as follows. The Mn—O bond is probably mainly ionic and Mn^{II} is quite readily accepted into the ionic Mg, Fe-silicate structures because the radius of Mn^{2+} is similar to the radii of Mg^{2+} and Fe^{2+}, and furthermore the Mn—S bond is sufficiently covalent and some Mn^{II} is accepted into many sulphide minerals, such as pyrite and sphalerite. It may be noted, too, that MnS is dimorphic, as laboratory experiments have shown that this compound may be either structurally ionic (NaCl) or covalent (ZnS). It seems, therefore, that in more senses than one, Mn^{II} is distinctly a "borderline" example. All elements on the right of Mn either form sulphide minerals of their own or enter into the structures of other sulphide minerals. As for the univalent cations, electronegativity values support conclusions from the ionization potential data, but again the picture is not nearly so clear.

For the large divalent ions the dividing line for the second ionization potential is about 14.5 - 15.0 eV. Tin is usually quadrivalent, but if divalent, may evidently enter sulphide minerals. All elements on the right of tin, form, with sulphur, bonds which are of such covalency that they either form minerals of their own or are acceptable into the structures of some other sulphide minerals.

Some observations on the behaviour trivalent metals have been made by Ahrens (1964).

SECTION (iii)

When the author commenced writing on element distribution in igneous rocks and their constituent minerals, it soon became apparent that an article of monograph size would have to be written in order to give adequate recognition to each of the many factors (see below), several of them interrelated, that are involved. Also, many viewpoints have been expressed in which emphasis is placed on one (or more) properties and in some publications *IP* is not considered at all. There is the question also as to the importance of utilizing thermodynamics.

It will be assumed that bonding in the common silicate minerals is ionic because coordination of oxygen about the different metals (Na, K, Ca, Fe, Mg, Al and Si) is in accordance with the ionic radius ratio rules. It has been suggested that the Si—O bonds in the SiO_4 tetrahedra are due to sp^3 hybridization and is therefore covalent. A question immediately arises - can such hybrids coexist in silicate structures where bonds between oxygen and Na, K, Ca, Fe^{II}, Mg and Al are evidently electrostatic?

The number of factors which influence element distribution are many and include: size (ionic radius), ionic charge, and polarizing power as indicated by the *n*th *IP*; electronegativity; order of arrival at site in a crystal growing from a melt; coordination and stereochemistry; for example, tetrahedral and in some rare instances octahedral coordination of oxygen about Zn and coplanar bond distribution as shown by Cu^{II}, Pt^{II} and Pd^{II}; crystal field stabilization involving *d*-transition elements with incomplete shells or half-shells, Cr^{3+} (d^3) and Ni^{2+} (d^8), for example; polarizability of the cation particularly if it is singly charged and large Cs^+ for example; polarizability of O^{2-}; formation of "complexes" such as $SiO_4{}^{2-}$ in the liquid phase; the nature of the bond both in the liquid and solid phases; the effect of the presence of accessory sulphides; halogen transport; the degree of covalency in the metal-oxygen bond as determined mainly by the polarizing power of the cation and the polarizability of the anion; and, for trace metal distribution, the structure of the host mineral (structural control).

In order to assess the importance of *IP* it is desirable to study the behaviour of pairs of cations which have constant charge and similar radii. The general rule is that for cations of given charge and similar radius, covalency of the bond with a non-metal will increase with increasing ionizational potential. For isostructural minerals and compounds, covalency gives rise to a lower heat of formation, a lower melting point and an increase in refractive index.

K^+(1.33, 4.34) - Rb(1.45, 4.18) - Cs(1.67, 3.89)

Both Rb^+ and Cs^+ substitute for K^+ in potassium minerals (feldspar and mica). The K—Rb association is close and that of K—Cs less so. These facts are well established and the relatively smaller coherence of K—Cs is evidently due to a relatively large radius difference. In fact it seems likely that the K—Rb—Cs association is controlled largely, if not completely, by ionic radius and the structures of the host minerals, K-feldspar and mica. The Rb—O and particularly the Cs—O bonds may be weaker than that of K—O, but these differences can be ascribed to length of bond and not to small *IP* differences. It should be borne in mind that Cs^+ is easily polarized and, if present, its polarization could have a covalency-weakening effect.

K^+, Rb^+ - Tl^+(1.47, 6.1) - Ag^+(1.27, 7.58) - Au(1.37, 9.23)

K, Rb - Tl

Soon after its discovery by Sir Wm. Crookes in 1861, Dumas dubbed Tl as the duckbill platypus (the ornithorhynchus) among the elements because of its dual chemical behaviour. Thus, Tl^I can behave as if it were a heavy metal similar to Ag^I, on the one hand, and on the other, mimics the heavy alkali metals, K and Rb. Geochemically, Tl also exhibits dualistic tendencies; traces of Tl are found in many sulphides and in all K-minerals. This accords with the relatively high *IP* of Tl, but what is not clear is to what extent the Tl—O bond in silicates is affected by *IP*.

As Rb and Tl have near-identical radii it is of considerable interest to examine their relationship in igneous rocks and minerals. In rocks the association is evident but not close and may be disturbed by the presence of sulphides and transport of volatile Tl - halides. In minerals, however, the relationship is quite remarkable, as shown by the investigation of Ahrens (1948) who observed a close association in a large variety of silicate minerals, including relatively rare amazonite and lepidolite, minerals which are of late formation in pegmatites. Although the Tl—O bond is evidently weaker than that of Rb—O in many compounds (heat and melting data) due to fairly high *IP* and a consequent covalency effect, possible geochemical significance is not apparent with regard to the K-minerals. An enrichment of Tl relative to Rb would be expected in the products formed by late-stage processes in pegmatites, but this is not so as the ratio, Rb/Tl, in lepidolites - minerals in which the amounts of Rb and Tl are at a maximum - is similar to all the others over a 1000-fold range of concentrations; see Fig. 1 of Ahrens (1948). The Rb—Tl relationship has been studied in detail by Ahrens (1964).

K - Ag - Au

Unlike Tl, where the *IP* is moderately high, the *IP*s of Ag and particularly Au are very high.

On the basis of size, some authors have suggested that Ag^+ could substitute for K^+ in K-minerals. However, in line with its high *IP*, bonding of Ag to O in compounds is distinctly covalency-weakened (heat and melting data). Ahrens (1964) has stated that, "Two factors work against a strong K—Ag association - the stereochemistry of Ag (low co-ordination of O about Ag) and a high degree of covalency in the Ag-O bond." The possible presence of Ag in sulphides of igneous rocks could disturb the K-Ag relationship.

The *IP* of Au is so high that if formed, Au—O bonds would be very weak indeed. The properties of Au^I compounds leaves little doubt that the degree of covalency in them is very high indeed and greater than that in Ag compounds, and as coordination of oxygen - or any ligand for that matter - is never ionic in gold compounds, the likelihood of any K^+—Au^+ replacement and hence Au—K association, if any, is very slight indeed.

Na^+(0.97, 5.14) - Cu^+(0.96, 7.73)

The *IP* difference is large and Cu^I—O bonds are essentially covalent. The comments above on K and Ag and Au apply here also.

Mg^{2+}(0.66, 15.04) - Zn^{2+}(0.69, 17.96)

The geochemical distribution of Zn in silicates is complicated but nevertheless of much interest. This arises from the fact that the degree of covalency in Zn—O bonds is quite high because of a high *IP* and because coordination of oxygen about zinc may be either four-fold (tetrahedral) or six-fold (octahedral).

The geochemical association of Zn and Mg + Fe^{II} has been extensively examined by many workers; for references see Gurney and Ahrens (1973). Some workers have stressed the importance of stereochemistry and others have, in addition, emphasized the importance of the nature of the bond and the structure and composition of the host mineral. Thus, Zn is enriched in biotite, hornblende and amphibole relative to olivine. The first three minerals contain some fluorine and H_2O and as a result the Zn-ligand bonds in them tend to be more ionic than Zn—O in olivine.

As the Zn—O bonds should in general be weaker than those of Mg—O and Fe—O (*IP* data, supported by heat and melting data) zinc should enrich relative to Mg + Fe^{II} in differentiated suites of rocks and their constituent Mg + Fe^{II} minerals. A few data appear to support this relationship (see, for example, those of Gerasimovsky and Nesmeyanova cited by Ahrens, 1964).

According to Gurney and Ahrens (1973) the Zn concentration in the basalt samples they analysed is relatively uniform. (Such behaviour contrasts with what has been observed on Cd - see below.) To what extent the nature of the Zn—O bond is important in this regard is difficult to ascertain.

Ca^{2+}(1.01, 11.87) - Cd^{2+}(0.97, 16.9)

As in the pair Mg—Zn, the *IP* difference is considerable and there is little doubt that in accordance with the supporting heat, melting and refractive index data, that the Cd—O bonds in oxygen compounds are distinctly more covalent and weaker than those of Ca—O in structurally similar compounds.

The geochemical consequences of the above observations are not clear. What is clear, however, is that the Ca—Cd association is barely evident; some of the data are contradictory. It seems likely that three factors act against a possible close Ca—Cd association: (i) the high *IP* of Cd and consequent covalence in the Cd—O bond, (ii) a difference in stereochemistry, and (iii) the chalcophile tendency of Cd but not of Ca. With regard to (iii) Brooks and Ahrens (1961) noted that the Cd concentration varied considerably (X 25) in the dolerite (diabase) specimens they examined and stated that this was probably due to the chalcophile tendency of Cd. The amount of Ca in the same rocks varied only slightly.

Ca—Hg^{2+}(1.12, 18.76)

The second *IP* of Hg is very high indeed, and it is not surprising, therefore, that Hg—O bonds in oxygen compounds tend to be strongly covalent and weak, as indicated by measured data (heat, etc.). There appears to be no evidence for a Ca—Hg association in igneous rocks and their minerals and the Ca—Hg relationship is similar therefore to that of Na—Cu^{I} and K—Ag, Au.

Mg—Fe(0.74, 16.18)

The transition metal Fe has been included because it is an abundant and petrologically important element and because crystal-field stabilization is relatively small, compared with Ni, for example, in which case *IP* rather than crystal field stabilization may be a significant factor.

A considerable amount of information is available on the petrological and geochemical relationship between Mg and Fe in igneous rocks and their minerals. Fe^{2+} and Mg^{2+} substitute, one for the other, freely in mafic (ferromagnesian) minerals and a characteristic feature of their relationship is that the ratio Mg/Fe is at a maximum in such minerals of early crystallization and falls steadily as differentiation proceeds. The strength of the Fe—O bond is distinctly less and more covalent than that of Mg—O as indicated by several measurements on isostructural minerals - refractive indices (Fe-rich minerals > Mg-rich ones) and, particularly, melting points. For example, the melting point of Mg_2SiO_4 (forsterite) is 1890°C compared with that of 1220°C for Fe_2SiO_4 (fayalite). It seems likely, therefore, that the higher *IP* of Fe relative to that of Mg has played a significant role, but the fact that the length of the Fe—O bond > the length of the Mg—O may also be significant on the assumption that the longer the bond, the weaker it is.

Al^{3+}(0.51, 28.45) - Ga^{3+}(0.57, 30.71)

A vast amount of information on the Al–Ga relationship is available and it is clear that the association between these two Group 2 elements, one abundant and the other rare, is close, as originally indicated by the early research of V. M. Goldschmidt and his co-workers in the 1930s. Willis (1978) has reviewed the data, lists several publications, and for the purpose of examining the "fine structure" of the Al–Ga relationship both in terrestrial rocks and in meteorites, gives many new data of high quality.

According to the *IP* values, the Ga–O bond should be more covalent and weaker than that of Al–O in respective isostructural compounds and this is supported by measured data-refractive indices, heats of formation and melting points (Ahrens, 1964). Accordingly, a slight but distinct enrichment of Ga relative to Al is anticipated during the process of differentiation. This appears to hold for some suites of rocks and not for others.

There appears to be a distinct tendency for Ga to enrich relative to Al in some aluminous minerals from pegmatites, amazonite (green microcline), for example, and particularly lepidolite; both minerals are of late formation in pegmatites.

In his 1964 paper Ahrens wrote: "The above slight but distinct tendencies accord with the fact that the Ga–O bond is slightly weaker than the Al–O bond; but it is not altogether clear as to which of two possibilities is the principal contributing cause - weakening due to the greater length of the Ga–O bond or weakening due to greater covalency in the bond. The second alternative is perhaps more important. Unfortunately other complications of interpretation may arise. In the first place, whereas both Al^{3+} and Ga^{3+} may assume tetrahedral or octahedral coordination, the tendency for Ga^{3+} to be octahedrally coordinated is presumably greater because of its greater radius. Secondly, Ga^{3+} may to some extent replace Fe^{3+}, as in magnetite."

According to Shaw (1957) the relative affinity for Ga in minerals from igneous rocks is, muscovite > plagioclase > biotite > hornblende > potash feldspar. The possible significance of slight differences in the character of the Ga–O and Al–O bonds in the above "affinity series" has not been studied.

Sc^{3+}(0.81, 24.76) - In^{3+}(0.81, 28.03)

Although the amount of analytical data is very considerable, particularly on Sc, it seems difficult to ascertain to what extent the *IP* difference of some 3 eV is geochemically significant.

The *IP* data support the fact that In but not Sc can be chalcophile and Shaw has suggested that the indium of igneous rocks may be concentrated in accessory sulphides, but Wager, Smit and Irving could find no evidence of this (for references, see Ahrens, 1964).

According to the *IP* data, the In–O bond in In_2O_3 should be more covalent and weaker than that of Sc–O in Sc_2O_3 and the heats of formation data, ΔH = 223 kcal/mole and 407 kcal/mole respectively support this prediction. As the radii of Sc^{3+} and In^{3+} are similar to Fe^{2+} and Mg^{2+}, Sc and In tend to find a home in pyroxene and according to the above heat data, the In–O bond should be weaker than that of Sc—O in this mineral. Consequently In should tend to enrich relative to Sc in differentiated rocks and minerals. Some evidence on rocks from the well-known Skaergaard intrusion support this (Wager, Smit and Irving) but in any event the Sc—In association is not close, possibly due to a fairly high degree of covalence in the In—O bond.

Rare Earth elements (Y and the lanthanons)

The characteristic oxidation state of these elements, with the exception of Eu, is three in a geological environment; Eu tends to occur as Eu^{II}. The radii of R^{3+} ions vary from 0.85 Å to 1.06 Å, and the third *IP* from 19 to 25 eV.

Taken as a whole the chemical properties of these elements are similar and as a consequence the Rare Earths form a geochemically coherent group. Nevertheless, a variety of distribution patterns and trends have been observed which are due to one or more properties of the R^{3+} ions. As the radius range is quite large, this property undoubtedly is a major factor in determining distribution patterns either in specific minerals or in suites of differentiated rocks. The possible significance of *IP* is, however, not clear.

Whereas the radii of R^{3+} cations decrease regularly with atomic number from La to Lu, that of the third *IP* is not as regular and is more complicated. Thus, for example, the third *IP* of Lu (20.96 eV) is less than the third *IP*s of most Rare Earth elements. The *IP* range of some 6 eV is quite considerable and it seems therefore possible that bonding in some REE—O silicates may be somewhat covalent and geochemically significant. This possibility has not been examined and appears worthy of a detailed investigation, particularly as a large number of highly quantitative analytical data are available.

Rare Earths - Bi^{3+}(0.96, 25.56 eV)

Despite different electronic structures, the chemical behaviour of Bi^{III} is similar to that of the Rare Earth elements in some respects. The relationship is, however, not as close as that between Tl^{I} and K + Rb where electronic structures also differ. Measured (heat, etc.) data support the *IP* data that the Bi—O bond in R_2O_3 oxides is distinctly more covalent and weaker than the Rare $Earth^{III}$—O bonds. The association of Bi and the Rare Earths is not therefore expected to be close. Distinct traces of Bi have been found in the Rare Earth mineral, gadolinite (Gurney and Ahrens 1973, who list several earlier publications), but otherwise geochemical association appears to be poor, as in suites of differentiated igneous rocks, and is probably due to two reasons: covalency, and the chalcophile tendency of Bi as a result of which Bi may be concentrated in accessory sulphides of igneous rocks.

Zr^{4+}(0.79, 34.34) - Hf(0.79, 33.3)

The close chemical and geochemical similarity between Zr and Hf is well established and has been known for a considerable time. Similar - possibly identical-radii; similar though not identical crystal chemistry and near-identical *IP*s are the main reasons for such similarity in the properties of Zr and Hf. Yet, the chemical and geochemical relationship is not perfect as small but distinct differences exist.

In a previous work (Ahrens, 1964), the fourth *IP* of Hf was listed as approximately 31 eV and it was concluded therefore that the Zr—O bonds in oxy-compounds would be slightly but distinctly more covalent than that of Hf—O bonds. Supporting evidence is as follows:

(i) basicity; that of $HfO_2 > ZrO_2$;
(ii) refractive index; that of some oxygen and fluorine containing compounds of Zr slightly > that of isostructural Hf compounds;
(iii) melting-point; that of ZrO_2 (2700°C) slightly less than that of HfO_2 (2780°C);
(iv) heat of formation; that of ZrO_2 (ΔH = 260 kcal/mole) slightly less than that of HfO_2 (ΔH = 266 kcal/mole).

According to current data the *IP* difference is small (1 eV), but nevertheless that of Zr remains > that of Hf and therefore supports the above relationships. Accordingly, Zr should tend to increase relative to Hf during differentiation of igneous rocks. If anything the reverse tends to happen: perhaps crystal chemistry may be a more important factor than small bond differences.

Transition metals

Transition metals with incomplete shells and half-shells have with the exception of Fe been excluded because it is difficult to establish to what extent crystal field stabilization contributes to the strength of the bond. A few brief notes will, however, be made on the distribution of Ni^{II} (d^8s^2) and Cr^{III} (d^3s^2) because their behaviour is characteristic and striking; see, for example, Ahrens (1964).

Both elements are strongly enriched in the earliest products (rocks and constituent Mg—Fe minerals, olivine and pyroxene) of differentiation and concentration decreases sharply with differentiation index. In an attempt to explain the striking behaviour of Ni and Cr, V. M. Goldschmidt had already in the 1930s referred to "special forces" which were evidently associated with the cations Ni^{2+} and Cr^{3+}. In this general respect it may be noted that the *IPs* are high and that crystal field stabilization is at a maximum. The importance of crystal field stabilization for geochemistry has been discussed in some detail by Curtiss (1964).

SECTION (iv)

During the weathering of rocks and minerals, the elements are redistributed and their fate depends on many factors. Some elements are redeposited *in situ* but most are transported by rivers and enter lakes and the oceans which it may be recalled make up 75% of the earth's surface. Their residence time in the oceans depends on many controlling factors which for our purpose need not be considered here. Detailed information on the residence times of fifteen elements is given by Goldberg and Arrhenius (1958). The time ranges from 1.0×10^2 years for Al, to 2.6×10^8 years for Na. Eventually a high proportion of the elements form minerals and find a home as sediments of the deep oceans which ultimately consolidate to form sedimentary rocks, the most abundant of which is shale. These rocks are composed of two major fractions, one inorganic and which is composed mainly of clay minerals and the other an organic fraction which may comprise a percent or so of many shales.

The unconsolidated sediments of the deep oceans contain numerous organic substances including some twenty amino acids which, as noted above, form stable complexes with several metals. Enrichment of metals, notably those with high ionization potentials, may therefore enrich in these ocean bottom sediments.

The organic matter of shales is enriched in many metals including Ag, Co, Cu, Mo, Ni, Pb, Sn, V and Sn. A variety of processes cause such enrichment including the formation of metallo-organic complexes. Ahrens (1966) has indicated the importance of the ionization potential with respect to metal enrichment in the organic matter of shales. It is pointed out that charge and radius must also play a role as to whether enrichment takes place or not, and if so, the degree of enrichment. Although it appears difficult to quantitatively correlate enrichment with *IP* because of the several factors involved (charge, radius and *IP*) it is nevertheless clear that enrichment is confined to those metals which have high *IP*s. A closer scrutiny of the available data may reveal some quantitative or semi-quantitative correlations.

CHAPTER 17

Ionization Potentials and the Radii of Cations

The relationship between *IP*s and cationic radii has been examined by Ahrens (1952) and most of what follows is extracted from that article. As highly charged *cations* such as F^{7+}, Cl^{7+}, Br^{7+} and I^{7+} may appear to be a misnomer to some the term *core* may be regarded as synonymous. For example, in the isoelectronic sequence Na^{+}, Mg^{2+}, Al^{3+}, Si^{4+}, P^{5+}, S^{6+} and Cl^{7+}, most would agree that Na^{+} in a crystal of NaCl could be regarded as an ion, but disagreement would arise as to the status of Cl^{VII} in potassium perchlorate ($KClO_4$).

Two relationships involving constants and regular variations have been observed to hold between *IP* and radius:

(a) one for constant charge ($\mathcal{Z}$)

and (b) one for isoelectronic sequences; electronic structure of cation (or core) remains the same, as in the sequence $Na^{+} \rightarrow Cl^{7+}$, above.

Although (a) and (b) are separate in one sense, they are related in another sense because of the nature of the different progressive changes involving, r, *IP* and $\mathcal{Z}$, as shown in Fig. 1 of Ahrens (1952) and as indicated below.

Relationships involving (a) will be considered first.

For the 8-electron cations Na^{+}, K^{+}, Rb^{+} and Cs^{+}, and Mg^{2+}, Ca^{2+}, Sr^{2+} and Ba^{2+}, an inverse relationship

$$r \propto \frac{1}{I_n^2}$$

or rI_n^2 = K, has been observed and appears to be closely obeyed. Thus, for the alkali metals, K = 25 and using K^{+} = 1.33 Å (*IP* = 4.34 eV) as reference ion (for reasons, see Ahrens, 1952) the radii of Na^{+}, Rb^{+} and Cs^{+} calculate as 0.97 Å, 1.46 Å and 1.67 Å respectively when the relevant *IP*s are used. These radii are close to those which appear in published sets of radii. The 2-electron cations Li^{+} and Be^{2+} do not fit the above expression for the 8-electron ions. If the expression did hold for Li^{+}, its radius would calculate (*IP* = 5.39 eV) at about 0.85 Å which is much greater than a value of 0.68 Å, or similar. The singly and doubly charged 18-electron cations,

Cu^+ (97 Å), Ag^+ (1.26 Å) and Au^+ (1.37 Å) and Zn^{2+} (0.74 Å), Cd^{2+} (0.97 Å) and Hg^{2+} (1.10 Å), likewise do not fit the 8-electron expression. In this respect two features are clear: first, *IP*s of these elements are much greater than those of the singly and doubly charged 8-electron elements with cations of similar size magnitude: second, whereas radii increase with weight in the groups, Cu^+, Ag^+ and Au^+, and Zn^{2+}, Cd^{2+} and Hg^{2+}, the *IP*s of the heaviest elements, Au and Hg, are actually greater than those of the lighter members of each group. It is of interest to note, however (Fig. 1 of Ahrens, 1952), that as charge increases in the sequences $Li^+ \rightarrow F^{7+}$, $Cu^+ \rightarrow Br^{7+}$, $Ag^+ \rightarrow I^{7+}$ and $Au^+ \rightarrow Bi^{5+}$, the relative effects noted above (elevation of *IP*) become progressively smaller and an inverse-square relationship appears to restore itself. It is in this sense that (a) and (b) are interrelated.

Entry of 3*d* electrons apparently disturbs the inverse-square relationship. In the group, Al^{3+}, Sc^{3+}, Y^{3+} and La^{3+}, radius increases from 0.51 Å to 1.14 Å, but the third *IP* of Sc is apparently elevated for a cation of radius 0.81 Å. This feature has been referred to as a "scandinide" effect (Ahrens, 1952) and becomes progressively more marked with increase in charge and reaches a maximum at Mn^{VII}. This feature contrasts with the developments involving $Cu^+ \rightarrow Cl^{7+}$, etc., noted above.

With regard to (b), smooth inverse relationships involving each of the isoelectronic sequences, $Li^+ \rightarrow F^{7+}$, $Na^+ \rightarrow Cl^{7+}$, $K^+ \rightarrow Mn^{7+}$, $Rb^+ \rightarrow Mo^{6+}$, $Cu^+ \rightarrow Br^{7+}$ and $Au^+ \rightarrow Bi^{5+}$ are seemingly well developed (Fig. 1 of Ahrens, 1952) and are similar in form. Mathematical expressions have not been formulated.

Figure 17.1 relates the *n*th *IP* to the reciprocal of the cation (or core) radius for the isoelectronic sequence $Na^+ \rightarrow Cl^{7+}$. Taken as a whole, the relationship is well developed as are others (not shown) involving isoelectronic sequences, except that which begins with Cs^+, Ba^+, La^{3+} and is then disturbed by the entry of the 4*f* electrons (lanthanide contraction). As a result of this contraction radii in the pairs Zr^{4+} - Hf^{4+}, Nb^{5+} - Ta^{5+} and Mo^{6+} - W^{6+} are either identical or nearly so. According to the *IP* data available, the value of the lighter component of the pair is always greater than that of the heavier. The highly elevated *IP* of Au has been noted as a consequence of the entry of the 4*f* electrons.

A final comment - the author is of the opinion that the *r*, *IP*, charge relationship should be re-examined as quantitatively as possible as the relationships appear too basic.

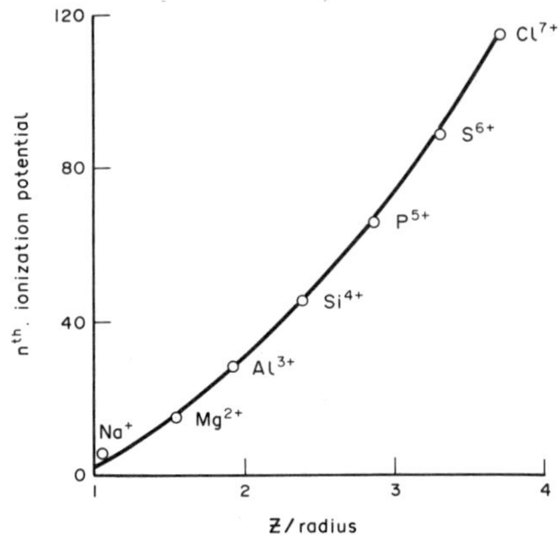

Fig. 17.1. The relationship between the *n*th ionization potential and Ƶ/ionic radius. This relationship appears to be well developed.

APPENDIX

Ionization Potentials*

as listed by Moore (1971)

Z	Element	I	II	III	IV	V	VI
1	H	13.598					
2	He	24.587	54.416				
3	Li	5.392	75.638	122.451			
4	Be	9.322	18.211	153.893	217.713		
5	B	8.298	25.154	37.930	259.368	340.217	
6	C	11.260	24.383	47.887	64.492	392.077	489.981
7	N	14.534	29.601	47.448	77.472	97.888	552.057
8	O	13.618	35.116	54.934	77.412	113.896	138.116
9	F	17.422	34.970	62.707	87.138	114.240	157.161
10	Ne	21.564	40.962	63.45	97.11	126.21	157.93
11	Na	5.139	47.286	71.64	98.91	138.39	172.15
12	Mg	7.646	15.035	80.143	109.24	141.26	186.50
13	Al	5.986	18.828	28.447	119.99	153.71	190.47
14	Si	8.151	16.345	33.492	45.141	166.77	205.05
15	P	10.486	19.725	30.18	51.37	65.023	220.43
16	S	10.360	23.33	34.83	47.30	72.68	88.049
17	Cl	12.967	23.81	39.61	53.46	67.8	97.03
18	Ar	15.759	27.629	40.74	59.81	75.02	91.007
19	K	4.341	31.625	45.72	60.91	82.66	100.0
20	Ca	6.113	11.871	50.908	67.10	84.41	108.78
21	Sc	6.54	12.80	24.76	73.47	91.66	111.1
22	Ti	6.82	13.58	27.491	43.266	99.22	119.36
23	V	6.74	14.65	29.310	46.707	65.23	128.12
24	Cr	6.766	16.50	30.96	49.1	69.3	90.56
25	Mn	7.435	15.640	33.667	51.2	72.4	95
26	Fe	7.870	16.18	30.651	54.8	75.0	99
27	Co	7.86	17.06	33.50	51.3	79.5	102
28	Ni	7.635	18.168	35.17	54.9	75.5	108
29	Cu	7.726	20.292	36.83	55.2	79.9	103
30	Zn	9.394	17.964	39.722	59.4	82.6	108

*1 cm^{-1} = 0.000 123 981 eV.

Appendix (*cont.*)

Z	Element	I	II	III	IV	V	VI
31	Ga	5.999	20.51	30.71	64		
32	Ge	7.899	15.934	34.22	45.71	93.5	
33	As	9.81	18.633	28.351	50.13	62.63	127.6
34	Sc	9.752	21.19	30.820	42.944	68.3	81.70
35	Br	11.814	21.8	36	47.3	59.7	88.6
36	Kr	13.999	24.359	36.95	52.5	64.7	78.5
37	Rb	4.177	27.28	40	52.6	71.0	84.4
38	Sr	5.695	11.030	43.6	57	71.6	90.8
39	Y	6.38	12.24	20.52	61.8	77.0	93.0
40	Zr	6.84	13.13	22.99	34.34	81.5	
41	Nb	6.88	14.32	25.04	38.3	50.55	102.6
42	Mo	7.099	16.15	27.16	46.4	61.2	68
43	Tc	7.28	15.26	29.54			
44	Ru	7.37	16.76	28.47			
45	Rh	7.46	18.08	31.06			
46	Pd	8.34	19.43	32.93			
47	Ag	7.576	21.49	34.83			
48	Cd	8.993	16.908	37.48			
49	In	5.786	18.869	28.03	54		
50	Sn	7.344	14.632	30.502	40.734	72.28	
51	Sb	8.641	16.53	25.3	44.2	56	108
52	Te	9.009	18.6	27.96	37.41	58.75	70.7
53	I	10.451	19.131	33			
54	Xe	12.130	21.21	32.1			
55	Cs	3.894	25.1				
56	Ba	5.212	10.004				
57	La	5.577	11.06	19.175			
58	Ce	5.47	10.85	20.20	36.72		
59	Pr	5.42	10.55	21.62	38.95	57.45	
60	Nd	5.49	10.72				
61	Pm	5.55	10.90				
62	Sm	5.63	11.07				
63	Eu	5.67	11.25				
64	Gd	6.14	12.1				
65	Tb	5.85	11.52				
66	Dy	5.93	11.67				
67	Ho	6.02	11.80				
68	Er	6.10	11.93				
69	Tm	6.18	12.05	23.71			
70	Yb	6.254	12.17	25.2			
71	Lu	5.426	13.9				
72	Hf	7.0	14.9	23.3	33.3		
73	Ta	7.89					
74	W	7.98					
75	Re	7.88					
76	Os	8.7					
77	Ir	9.1					
78	Pt	9.0	18.563				
79	Au	9.225	20.5				
80	Hg	10.437	18.756	34.2			
81	Tl	6.108	20.428	29.83			
82	Pb	7.416	15.032	31.937	42.32	68.8	
83	Bi	7.289	16.69	25.56	45.3	56.0	88.3
84	Po	8.42					

Appendix (*cont.*)

Z	Element	I	II	III	IV	V	VI
85	At						
86	Ra	10.748					
87	Fr						
88	Ra	5.279	10.147				
89	Ac	6.9	12.1				
90	Th		11.5	20.0	28.8		
91	Pa						
92	U						
93	Np						
94	Pu	5.8					
95	Am	6.0					

BIBLIOGRAPHY

Ahrens, L. H. (1948) The unique association of thallium and rubidium in minerals. *J. Geol.* 56, 578.

Ahrens, L. H. (1952) The use of ionization potentials. Part 1. Ionic radii of the elements. *Geochim. Cosmochim. Acta* 2, 155.

Ahrens, L. H. (1953) The use of ionization potentials. Part 2. *Geochem. Cosmochim. Acta* 3, 1.

Ahrens, L. H. (1954) Shielding efficiency of cations. *Nature* 174, 644.

Ahrens, L. H. (1956) Some ionization potential variations and relationships. *J. Inorg. Nucl. Chem.* 2, 290.

Ahrens, L. H. (1957) Some ionization potential variations and relationships. II. *J. Inorg. Nucl. Chem.* 4, 264.

Ahrens, L. H. (1958) Variation of refractive index with ionization potential in some isostructural crystals. *Mineralog. Mag.* XXXI, 929.

Ahrens, L. H. and Morris, D. F. C. (1956) Ionization potentials and the chemical binding and structure of simple cristalline compounds : II. *J. Inorg. Nucl. Chem.* 3, 270.

Ahrens, L. H. and Taylor, S. R. (1961) *Spectrochemical Analysis*, 2nd edition. Addison-Wesley, Pergamon Press.

Basolo, F. and Pearson, R. G. (1958) *Mechanism of Inorganic Reactions*. John Wiley.

Bernal, J. D. and Megaw, H. D. (1935) *Proc. Roy. Soc.* A 151, 385.

Bjerrum, J., Schwarzenbach, G. and Sillen, L. G. (1957) *Stability Constants. Part 1. Organic Ligands*. Special Publication No. 6 of the Chemical Society, London.

Boumans, P. W. J. M. (1966) *The Theory of Spectrochemical Excitation.* Hilger & Watts, Ltd., London.

Brooks, R. R. and Ahrens, L. H. (1961) Some observations on the distribution of thallium, cadmium and bismuth in silicate rocks and the significance of covalency on their degree of association with other elements. *Geochim. Cosmochim. Acta* 23, 100.

Brown, C. B. (1956) Have we abandoned the physical theory of nature? *Progress* XLIV, 619.

Cartledge, G. H. (1928) Studies on the periodic system. *J. Amer. Chem. Soc.* 50, 2855.

Cartledge, G. H. (1930) Studies on the periodic system. *J. Amer. Chem. Soc.* 52, 3076.

Cotton, F. Albert and Wilkinson, Geoffrey (1966) *Advanced Inorganic Chemistry*, 2nd edition. Interscience.

Coulson, C. A. (1952) *Valence*. Oxford, at the Clarendon Press.

Cox, E. G., Wardlaw, W. and Webster, K. C. (1935) *J. Chem. Soc.* 1475.

Curtiss, C. D. (1964) Applications of crystal field theory to inclusion of trace transition elements in minerals during magmatic crystallization. *Geochim. Cosmochim. Acta* 28, 389.

Fersman, A. E. (1935) The EK-system. *Kokl. Akad. Nauk* 2.

Förland, T. and Weyl, W. A. (1949) Distribution of heavy metal ions in silicates containing several anions of different size. *J. Amer. Chem. Soc.* 32, 267.

Goldschmidt, V. M. (1926) Geochemische Verteilungsgesetze der Elemente. *Skrifter Norske Vid.-Akad., Mat.-Nat. Kl.*, No. 2.

Goldschmidt, V. M. (1954) *Geochemistry*. Oxford, at the Clarendon Press.

Gurney, J. J. and Ahrens, L. H. (1973) The zinc content of some ultramafic and basic rocks. *Tr. Geol. Soc. S. Africa* 76, 301.

Heslop, R. B. and Jones, K. (1976) *Inorganic Chemistry*. Elsevier, Amsterdam.

Huheey, James, E. (1968) *Inorganic Chemistry (Principles of Structure and Reactivity)*, 2nd edition. Harper & Row.

Irving, H. and Williams, R. J. P. (1953) The stability of transition-metal complexes. *J. Chem. Soc.* 637, 3192.

Kastner, S. O. (1969) Approximate higher ionization potentials for the heavier elements. *J. Chem. Phys.* 50 (7), 2896-109.

Lakatos, B., Bohus, J. and Medgyesi, Gy. (1959) A new way for the calculation of the degree of polarity of chemical bonds. II. *Acta Chim. Hung.* 20, 1.

Martin, W. C., Hogan, Lucy, Reader, Joseph and Sugar, Jack (1974) Ground levels and ionization potentials for lanthanide and actinide atoms and ions. *J. Phys. Chem.* Ref. data Vol.3, No.3.

Moore, C. E. *Ionization Potentials and Ionization Limits derived from the Analyses of Optical Spectra*. NSRDS - NBS 34, National Bureau of Standards, Washington, D.C., U.S.A.

Poincaré, H. (1929) *The Foundations of Science*. The Science Press, New York.

Pyper, N. C. and Grant, I. P. (1978) The relation between successive atomic ionization potentials. *Proc. R. Soc. Lond.* A,339, 525.

Ramberg, H. (1953) Relationships between heats of reactions among solids and properties of the constituent ions, and some geochemical implications. *J. Geol.* 61, 318.

Shaw, D. M. (1957) The geochemistry of Ga, In and Tl. *Phys. Chem. Earth.* 2, 164.

Szádeczky-Kardoss, E. (1954) Vorläufiges über Anionpotentiale und Verbindungspotentiale. *Acta Geol. (Hungary)* 2 (3-4), 285.

Szadeczky-Kardoss, E. (1955) Das Verbindungs potentiale und seine Beziehungen zum Schmelzpunkt und zur Härte. *Acta Geol. (Hungary)* 3 (1-3), 115.

Vincent, E. A. and Crocket, J. H. (1960) Studies on the geochemistry of gold in rocks and minerals of the Skaergaard intrusion, East Greenland. *Geochim. Cosmochim. Acta* 18, 130.

Wells, A. F. (1962) *Structural Inorganic Chemistry*, 3rd edition, Oxford, at the Clarendon Press.

Willis, J. P. (1978) Some aspects of the geochemistry of gallium in silicate rocks and stony meteorites. Ph.D. thesis, University of Cape Town.

Wyckoff, R. W. G. (1951) *Crystal Structures*. Interscience.

Author Index